CW01474907

The Future of Seeing

# THE FUTURE
# OF SEEING

## How Imaging Is Changing
## Our World

## Daniel K. Sodickson

Columbia University Press
New York

Columbia University Press
*Publishers Since 1893*
New York    Chichester, West Sussex
cup.columbia.edu

Copyright © 2025 Daniel K. Sodickson
All rights reserved

Library of Congress Cataloging-in-Publication Data
Names: Sodickson, Daniel K. author
Title: The future of seeing : how imaging is changing our world /
Daniel K. Sodickson.
Description: New York : Columbia University Press, [2025] |
Includes bibliographical references and index.
Identifiers: LCCN 2025003920 | ISBN 9780231209922 hardback |
ISBN 9780231558181 ebook
Subjects: LCSH: Imaging systems—History | Imaging systems—
Technological Innovations—Forecasting | Imaging systems—
Social aspects | Vision
Classification: LCC TK8315 .S635 2025 | DDC 621.36/709—dc23/eng/20250327
LC record available at https://lccn.loc.gov/2025003920

Printed in the United States of America

Cover design: Henry Sene Yee
Cover image: K. H. Fung/Science Photo Library

GPSR Authorized Representative: Easy Access System Europe,
Mustamäe tee 50, 10621 Tallinn, Estonia, gpsr.requests@easproject.com

To my father, Lester, who taught me how to ask questions about the world.

To my mother, Isabel, who taught me how to ask questions about people.

To my nuclear and magnetic family, Sarah, Hannah, and Noah, who taught me that the answers are always more than they appear.

# Contents

# Prologue: An Imager's Perspective

Sometimes, when I'm walking through Manhattan's familiar streets, a sudden shift of perspective jolts me. No longer do I register the comings and goings of vehicles and pedestrians. The eye-catching messages of billboards and bus-stop schedules slip into obscurity. All the everyday particulars of life in a busy city fall away. In their place, I see a world bathed in light.

When I am in the clutches of this perspective, it is not the things I see that strike me. It is the means by which I see them. Light—the fastest messenger in the universe—is shuttling back and forth between the shadowed surfaces of things. Light bounces, buzzes, zips, and whizzes. It snoops into dark corners, airing dirty laundry faster than a thought. It fills the air with invisible energy, mingling umbers and tans with olives and golds, making everything it touches twinkle and gleam. Or else, on dimmer days, light softly settles like a muffling blanket of snow. The light is permeating, general, a quiet sheen of ambient illumination. It settles softly over everything, bringing me news of all the living and the dead.

I am an imager, you see. I design MRI machines and other devices that take pictures of what light cannot reach. The way I view the world is conditioned by a quarter-century of studying how images are made. And oh, wonder! What marvelous images can be made in today's world. These images do not merely reflect the surface of things. They reach inside as well.

When I look closely enough at images rendered on-screen by the medical imaging machines that have been central to my professional life for more than twenty-five years now, something else emerges. I see a record of interactions with the intimate interior of a living body. Through this translucent script, the body speaks. It tells its story.

This is the story I would like to tell you now. It is the story of how we humans came to see ourselves in remarkable new ways. It is the story of still newer ways of seeing that will make today's most eye-catching images seem like faded sepia snapshots. By the time the story ends, I hope you will have begun to see things in a new light.

I am an imager, and, whether or not you know it yet, so are you. This is our story.

The Future of Seeing

# Introduction

## *The Story of Imaging*

**W**hat if you could pop into a drugstore for an MRI?

What if your clothes could look into your internal organs to confirm you don't have cancer?

What if ubiquitous surveillance cameras could read your mind?

These scenarios sound like the stuff of science fiction. As much as I enjoy thoughtful science fiction, though, fiction is not my primary concern here. I pose these questions to you, dear reader, by way of introduction to an age-old but also very modern conundrum: the conundrum of seeing.

Seeing is at the core of how many of us understand our world. "Seeing is believing," as the saying goes, or "What you see is what you get." We humans have many means of interacting with the world around us, and vision in the strict biological sense is not essential to our functioning, particularly in a world with ever-advancing technologies for nonvisual communication and even sensory enhancement. The blind community has managed just fine without relying upon mere eyesight, thank you very much. That said, we are a species that has long leaned heavily on its visual sense, and a broader conception of vision has come to be synonymous with understanding: "I see" is another way of saying "I understand."

The conundrum for all of us, then, is that the *way* we see is changing. The tools of modern astronomy enable us to extend our vision ever farther into outer space while modern microscopes and medical imaging devices

1

allow us to delve ever deeper into inner space. Using the pervasive tools of consumer electronics, we can record what we see and send that recording careening around the world at the speed of light to share with anyone and everyone. At the same time, thanks to the hardware and software brought to us by today's media giants, we can each see what we want to see, or what others want us to see. The era of human advancement that brought us the accountability of instant replays and dashcams also opened a Pandora's box of personalized searches and deep fakes. Such tools influence our decisions and cement our biases, giving our personal fictions the veneer of truth.

Just as the way we see is changing, so is the way we are seen. Less and less in today's world escapes automated scrutiny. In many a modern city, a casual stroll to the corner store will be captured by any number of electronic eyes along the way. Much of the promise of smart homes is predicated on tracking our every move when even our closest neighbors and friends cannot see us. Our skin is no longer a barrier to inspection when we pass through airport scanners. As technology advances, even our brains are becoming increasingly open books. These changes are already affecting our health, our privacy, and our relationships—and the pace of change is only accelerating. Since seeing, in its wider senses, is also at the core of how we understand ourselves, it is even fair to say that our sense of who we are is poised to change.

Are you ready for the future of seeing?

In this book, I explore the changing nature of seeing through the lens of imaging. Broadly speaking, *imaging* may be defined as the process of generating pictures, videos, or other representations of spatially organized information. *Imaging* encompasses all the natural and artificial methods we use to map the world around us—and within us. Many creatures who share the earth with us have evolved diverse imaging capabilities—some rudimentary, some remarkably advanced—that allow them to navigate their environments skillfully. Eagles are famous for their sharp eyes. Bats hunt their prey using "sound pictures." In addition to sharing some of these capabilities, humans have taken the artifice of imaging to great heights. What began with our unaided eyes has been transmuted into a triumph of technology—some of it familiar to the point of banality, some of it eye-crossingly arcane. We have created cell phone cameras, planet-girdling telescopes, and MRI machines.

Regardless of the form it takes, imaging is everywhere nowadays. From the ultrasound that may have peered in at you in utero to the cameras that increasingly record your public comings and goings and capture your private moments, you live an imaged life. Even the cherished memories that might once have resided only in your mind now take on new life in the form of images that you store in your camera and post online.

What of the future, then? The future holds a brave new world of imaging machines. The workings of these machines might seem as mysterious to most of us now as our current world of orbiting telescopes and buzzing medical imaging tubes might have seemed to a sixteenth-century glassmaker living in a time before humans had learned to build their first primitive spyglasses. Nevertheless, just as telescopes, microscopes, and cameras forever altered human history, these new imaging devices will help to determine the shape of our collective future.

The first purpose of this book is to prepare you for the changes that will be wrought on our lives by upcoming revolutions in imaging. To provide some perspective on these changes, I share with you a jaunty tale of previous imaging revolutions that brought us the fundamental methods we now use to make images.

I do not use the term *revolution* lightly. Imaging permeates the history of scientific revolutions, and societal revolutions are frequently tied to the way we see. The word *revolution* itself has its origins in descriptions of the motion of celestial bodies. And how do you think humankind first charted those celestial revolutions? With revolutionary imaging devices, of course.

While this book is concerned with the impact of imaging broadly defined, I often take the vantage point of medical imaging to illustrate this impact. Why? Because medical imaging is arguably the most intimate of human imaging approaches. To many, it is also one of the most forbidding. Whereas astronomy gives us tantalizing glimpses of vast spaces beyond our ken, medical imaging reaches inside us to uncover sometimes unsettling truths. It literally gets under our skin. In the worst of times, it is a bringer of bad tidings—the tool that doctors use to tell us we are sick.

But there is a great deal more to medical imaging than meets the eye. The devices that make up the medical imager's arsenal are not just physicians' tools—they are the outcome of a rich and remarkable history of invention that has forever changed the way we see and understand ourselves. These devices enable a diverse body of research into how our bodies—and any bodies—work. They are also closely connected, in their origins and operations, to the more familiar devices employed in photography and astronomy. Moreover, the very fact that medical imaging gets under our skin qualifies it as the perfect lens for investigating the future of seeing. Just as the eye-catching images of astronomy may lead us to contemplate our place in the universe, medical images raise for us all manner of quintessentially human questions: "Am I OK?" "Will my children be OK?" "How does my mind work?" "Who am I?"

Medical imaging is also a key catalyst for societal change. At the turn of the twentieth century, x-rays took the world by storm, opening up the body's secrets to inspection by specialists and nonspecialists alike, and creating entirely new modes of medical care. By the turn of the twenty-first century, cross-sectional imaging (i.e., the visualization of internal "slices" of the body using a wide range of approaches) had become a globe-spanning juggernaut driving multibillion-dollar industries, guiding the care of billions of patients, and placing substantial cost pressures on national economies. Likewise, today's emerging medical imaging and sensing technologies will change the face of health care, helping us transition from a time-honored but sluggish episodic model driven by symptoms to a model of continuous evaluation and preservation of each person's ongoing state of health. At the same time, our view of the human form, which has progressed from opaque to transparent to clinically dissected and disassembled, will soon be reanimated in dynamic, living color. Perhaps ironically, the means by which this revolutionary shift in perspective will take place promises to mimic the evolutionary processes by which we came to see in the first place.

Part I of this book looks back at the history of seeing. It draws a line from the early oceans, where organisms first developed eyes, to the present, where we use advanced imaging devices to slice through living bodies without making a single cut. Starting with the Precambrian Era and proceeding through the period of biological blossoming known as the Cambrian explosion about 540 million years ago, the storyline follows subsequent revolutions catalyzed by human-made imaging machines. I describe the invention of telescopes and microscopes in the 1600s, cameras in the 1800s, x-ray machines just before the turn of the twentieth century, and cross-sectional imaging devices like CT (or CAT) scanners and MRI machines in the 1970s. I recount the often surprising history of new technologies and clever transformations that have extended the reach of our vision to impossibly faraway places and unimaginably tiny things while also allowing us to see our own hearts beating or observe our minds at work. I connect stories of inventors, physicians, patients, and others caught up in the imaging enterprise with a progressive shift in how humans have come to view not only our expanding universe but also our increasingly visible bodies.

Part II looks forward, charting the emerging technologies and transformations that will change the way doctors see patients, the way people see the world around them, and the way all of us see ourselves. I explore how artificial intelligence and abundant sensors will allow us to emulate some of the deeper functions of our evolved biology, not merely imitating our eyes but emulating our brains and extended nervous systems. I show how these changes will jailbreak our imaging devices, bringing the benefits of imaging to people everywhere. I also consider tricky cultural and political implications of the democratization of imaging.

Apart from preparing you for the visible future bearing down on us, this book has another central purpose. For those who view the intimidating tubes of modern medical imaging—MRI machines, CT scanners, PET scanners, etc.—with understandable trepidation, I provide a more accessible and human-centric perspective on imaging. I describe some of the common origins, innovations, and ambitions that connect medical imaging with astronomy and microscopy, linking our ever-expanding view of outer space to the ongoing exploration of inner space. And I show how, in many startling ways, our imaging machines are built in our own image.

From the earliest dawn of human history, we humans have sought to extend our vision. Our distant ancestors looked up to the skies, wondering what mysteries lay behind the daily revolution of the sun. They huddled together around firepits, telling one another stories about the origins of starlight. Step by step, descendants of these first stargazers figured out how to build tools to help them see farther and more clearly. We are the inheritors of this legacy of seeing.

In the face of all-too-frequent reminders of the many forces that divide us, this quest for clarity unites us. The interpretation of what we see may have as many variations as there are beholders. The urge to see, though, precedes and underpins our interpretations, and it serves as a powerful metaphor for the restlessness of human curiosity. Whenever one extends one's vision, one invariably expands one's mind. The story of astronomy—of how we reached out to the farthest extent of the visible universe—has a broad and almost primal appeal. As the tools of our enhanced vision have advanced, they have been trained inward as well.

This brings me to a final modern conundrum of seeing. Imaging touches more lives in the modern world than it has ever done, yet its inner workings have become more mysterious than ever. Somewhere in the course of our burgeoning industrial revolutions, we delegated the understanding of enhanced vision to specialized experts. Images bring the news to life. They facilitate and illustrate scientific advances. They bring friends and family together across thousands of miles. But the mechanisms of imaging remain hidden.

To explore the future of seeing, I must first pull back this veil, and there is no better way to do so than to consider the history of seeing. I invite you, then, to come back with me to the beginning.

In the beginning, there was darkness. And then there was light.

# PART I

## LOOKING BACK

### The History of Imaging

# 1

## The Nature of Seeing

How eyes evolved, and how they became models for
human-made imaging approaches.

A pproximately 3.8 billion years ago, in the earth's early oceans, the
first cells were formed.

"OK, wait just a second!" I hear you cry. "What does imaging have
to do with the first cells?" In brief, the connection has quite a bit to do with
what an image actually is. For the sake of argument, let us define an image
as a representation of spatially organized information. When you look at
the *Mona Lisa*, you see a mysteriously amused figure in the foreground and
a much-debated otherworldly landscape in the background. Even though
all you are seeing is a flat sheet of canvas, your eyes and your brain automatically sort out the various components of the scene, just as they do when
you are crossing a busy street filled with potentially life-threatening vehicles
rushing back and forth. An image is a map of what is where in the world.

Now back to the early oceans. Keeping spatial information straight was
utterly essential to life in those primordial depths, just as it is today. As
the oceans began to fill with life, each patch of aquatic real estate was an
increasingly lively but also increasingly life-threatening environment. It
behooved an early sea creature to know that "food is over here." It also
behooved such a creature to know that "the hungry creature that thinks
*you* are food is over there."

The profound competitive advantage of knowing the difference between
here and there provided a powerful evolutionary impetus for early cells

9

and multicelled creatures. A biological arms race among the earliest denizens of Earth led to a profusion of new ways to sort out spatial information. One particularly successful evolutionary adaptation was a light-sensitive organ that could map the surrounding environment—an organ we now call an eye. We know from the fossil record that partway through the Cambrian Era, around 520 million years BCE, humble sea creatures called trilobites had developed remarkably complex organs that we now identify as compound eyes.

From the trilobite's compound eye, it was just a hop, skip and a jump to the human eye. A few evolutionary eras may have been required for refinement, but all the essential elements were there in the Lower Cambrian. Cambrian eyes already contained focusing devices, like our lenses, to collect light. They relied on light-sensitive molecules like the ones in our retinas. They used many adjacent light-detecting components, just as our retinas do, to distinguish location. One notable difference between trilobites and humans, of course, is that we managed to evolve one humdinger of an image-processing device: our sizable brain. In chapter 2, we will take up the question of what humans would come to accomplish using our brains. In the meantime, let's take a closer look at those first cells. Those cells represent our origins, but in them we may also discover the origins of imaging.

Earth's original oceans were full of material but devoid of life. The seas were repositories of all manner of molecular detritus, forming what the famous astronomer and scientific communicator Carl Sagan liked to call the "primordial soup." The base of this soup was, of course, water. Where did the water come from? It had escaped from a larger interior sea of molten rock and had condensed to cover the planet as it cooled.

Dispersed in all this water were any number of elements, including ions like sodium and chlorine, various catalytic metals, and assorted bits of carbon. All of it was piping hot. There was thermal energy aplenty in the simmering primordial soup. Atoms were free to mix, mingle, and combine into diverse molecules, everything tumbling along aimlessly whichever way the currents flowed. But when and how did any of that flotsam and jetsam begin to act as we expect living things to act, namely, with purpose?

What we might recognize as life began when molecules first learned the difference between *here* and *there*. As Sagan describes in the second episode of his well-known television series *Cosmos* (which kept me and countless other humans young and old spellbound starting back in 1980), a certain class of molecules called polar molecules were attracted to water at one end and repelled by it at the other end. As a result, they tended to clump together with their attractive heads all lined up to face the water on two sides, protecting their repulsive tails within their ranks. These little molecular regiments had a tendency to close ranks, forming spheres that contained pockets of seawater. And, lo and behold, there was an inside, and there was an outside. Cells were born.

Just one small step for a bit of flotsam was a giant leap forward for biological life. The vast symmetry of the primordial ocean was broken, and the seed of identity was sown. Now there was a *this* floating in the midst of all of *that*. A prototypical "I" came into being in a sea of otherness. Suddenly there was a *here* as opposed to a *there*. It is interesting to note that the word *paradise* is derived from early roots meaning "enclosure." If you like, you can think of each early cell as its own little Garden of Eden.

The simple binary distinction between *inside* and *outside* was a primitive form of spatial information. Knowledge is power, as they say, and this single bit of information turned out to hold remarkable potency. The ranks of aligned polar molecules, which we would now call membranes, didn't just part the waters; they eventually managed to change them. Suitable arrangements of membranes allowed conditions inside early organisms to differ from those in the external environment, first in the raw concentration of simple molecules, then in complex chemical composition, and eventually in detailed microstructure. Cells became receptacles for early genetic information, which evolved into RNA and DNA. Clusters of cells joined forces to form more intricate organisms. Organisms proliferated, colonizing the ocean.

This was all well and good. However, the blossoming of life came at a price. Primitive organisms needed energy to maintain their structural differentiation from the surrounding molecular soup. As we know from thermodynamics, energy is required to fight against the natural tendency of all things to slip into disorder. Sources of energy, therefore, became valuable, as did reserves of materials to replace anything that might be lost over time. Organisms needed resources to maintain their "thisness," in a process we now call homeostasis. The ability to seek out such valuable resources would have represented a substantial competitive edge for any being living in a world of aquatic drifters. For any sort of organized search, though, such a being would have needed a map.

Thus it was that organisms came to develop the ability to sense spatially resolved information about their surroundings. Pores and, in time, more advanced sensors allowed membranous entities to slide along chemical gradients, seeking out more of what they needed. Such entities could recognize one direction as more desirable than another. They could find their way to regions where nutrients were plentiful.

To this day, our bodies develop in this way. Coordinated changes in molecular signaling guide immature cells along complex paths that generate the intricate bio-weavings we call organs. Without knowing here from there, our cells would be utterly incapable of forming bodies. Spatial awareness is at the very root of life as we know it.

Spatial awareness is also a great advantage if one wants to avoid active threats to one's life. For Earth's early organisms, there was no more active threat than other hungry organisms. So early sea creatures learned to sense regions of potential danger. They learned to identify and avoid places where things were cold, dark, and ominous and where bigger creatures might be lurking.

They also developed new means of getting where they wanted to go. As cells and multicelled bodies became more complex, mechanisms of active locomotion evolved. Movement became deliberate, guided by spatial equivalents of the knowledge of right (i.e., what keeps me as I am) and wrong (i.e., what causes my essence to drain away or be gobbled up by somebody else). Organisms became hunters and gatherers. Predators stalked their prey. Prey hid. Life became an interactive, coordinated dance.

In such a dance, one needed more than ever to be aware of one's surroundings. So it was that our distant evolutionary ancestors traded up from rudimentary spatial awareness to what we now know as senses. Senses are the mechanisms by which living entities probe their environment in order to navigate through it safely and successfully. Chemical awareness—a prototypical sense of smell—may have been the first of the senses. As the Cambrian Era got going, though, a remarkable diversity of sensory organs and adaptations rapidly followed suit. Organisms evolved to use all manner of probes to map their immediate neighborhood. They came to use light (electromagnetic waves), sound (vibrational displacements), and just about any other means ready to hand. (Well, anything that was accessible and abundant, really, until structures like hands finally appeared on the scene.) As far as we can tell, Cambrian creatures were remarkably adept at sensing. In fact, the powerful competitive advantage conferred by senses like vision is sometimes credited with driving the Cambrian explosion in the first place.

In short, in climbing up the evolutionary ladder from the first cells, the world's early flora and fauna had learned a key lesson. They had learned to use the interactions of various probes with the world around them to yield spatially organized information about the structure of that world. This is not a bad working definition of *imaging*.

We have finally arrived, then, at the point of this little excursion into the origin of species. My point is that imaging is not a modern invention at all. Imaging originated in the sensing mechanisms that all successful species eventually developed over the course of their evolution. The quest to extend our vision is not merely natural—it is practically elemental.

From this point of view, imagers like me are in pretty good company. The early oceans were brimming with little imagers. This was the case even before those oceans began overflowing with Cambrian Era biodiversity, and it was certainly true afterward.

The living world today is even fuller of unwitting experts in the imaging arts. You are one. So is your cousin. So, too, is your cat. As you read this page, your eyes (or whichever set of senses you use for reading) are forming images of a sequence of characters and streaming those images directly to your brain for immediate processing. Your cat is forming images of its prey and planning its next move—perhaps a pounce into the lap of a certain foolish human sitting hunched over a book. My point is that, in one form or another, imaging is everywhere.

We give little thought to the innate mechanisms by which we all generate images as we go about our daily lives. These mechanisms are second nature to us. Our senses generally just deliver images to us, and we take it from there. But we can learn a lot from how our senses operate.

When it comes to imaging, one sense stands out, of course: the sense of sight. I will next describe some of the elementary physics and basic biology of vision. If neither physics nor biology is quite your thing, never fear. I will not delve too deeply, or for too long, because my real interest here is the origin story of imaging. For your part, you should feel free to skim as lightly as you choose, knowing that the bottom line is this: every feature of biological vision that I will describe in this chapter serves as a model for artificial forms of imaging that humans would eventually create.

Are you anywhere near a mirror right now? If you are, indulge me and go look yourself in the eye for a moment. (If no mirror is available but you have your cell phone handy, try holding the camera up to your face, removing your glasses if you have them, and flipping the view so that you can see yourself looking. If you are visually impaired, then just ignore me and probe the space around you with your other senses. Those senses will get their due when it comes to the future of seeing.[1])

I'll pause here to give you a little time for self-inspection . . .

OK, welcome back.

The eyes that you just saw are marvels of natural engineering. Yes, your eyes are truly spectacular. I am not trying to flatter you or make you uncomfortable. Neither am I extolling the expressive power of eyes nor celebrating the poetical window they may offer into the soul. I am speaking as a physicist and an engineer—a designer of tools. And, if I may say so, you have a truly beautiful pair of tools right there below your forehead.

Let us briefly take stock of what your eyes can do. First and foremost, they can detect the presence or absence of light. Open your eyes in the daytime, and light floods in like the wave we know it to be. At night, your eyes are sensitive enough to detect a single photon—the smallest quantal unit into which light can be divided. If there is light to be found, whether in overwhelming abundance or in solitary scarcity, your eyes will find it.

Second, your eyes are remarkably good at telling where that light is coming from. In addition to being avid collectors of light, each eye's fine-tuned and adjustable lens focuses incoming rays onto the precision detector array of your retina, creating a tiny replica of the world around you. When you take in an awe-inspiring landscape, your eyes set their sights on the infinite horizon, and the grand vista before you is literally copied into your head for you to record every detail in memory. If you wish to set your sights closer by, however, you can focus in on a page of this book, making out sequences of characters spaced a millimeter apart or inspecting the fine contours of an illustration.[2]

These two fundamental capabilities represent key features of any good imaging method: sensitivity and spatial discrimination.

In addition to these essential characteristics, your eyes have many other noteworthy capabilities. The light-sensitive molecules in the cells of your retina respond differently to different frequencies of light, *frequency* referring to the rate at which light waves oscillate from peak to trough to peak again. This allows you to see the world in color. Adding still more depth to your visual experience is the fact that you can judge distances between you and the things you see. Careful coordination between your two eyes, together with other delicate mechanisms of depth perception, allow your brain to experience the world in 3-D. In short, there is more to eyes than, well, meets the eye. The closer you look, the more ingenious mechanisms you will discover. Such multifaceted capabilities don't come for free, however. Your eyes are remarkably complex organs.

Human eyes are so complex, in fact, that they have fueled controversies ever since Darwin first proposed his theory of evolution. In *An Immense World*, a remarkable 2022 book on animal senses, as well as in an earlier article about animal eyes for *National Geographic*, Ed Yong succinctly summarizes these controversies and their potential resolution.[3] Allow me to summarize his summary for you quickly here.

Darwin himself wondered how intricate and delicately engineered structures such as the eyes could have evolved little by little from simpler elements. In *On the Origin of Species*, he wrote, "To suppose that the eye, with all its inimitable contrivances for adjusting the focus to different distances, for admitting different amounts of light, . . . could have been formed by natural selection, seems, I freely confess, absurd in the highest possible degree."[4] Evolution, you see, is blind. A small step on the way to an eye would need to confer some palpable advantage in its own right in order to be propagated along the evolutionary chain. There is no advantage to *almost* seeing. So how could a process driven by random mutations ever have the foresight to pull together all the elements needed for a fully functional eye?

In the same passage from *Origin of Species*, Darwin proposed an empirical test for whether natural selection was a plausible explanation for "organs of extreme perfection and complication" such as eyes. "Reason tells me," he continued, "that if numerous gradations from a perfect and complex eye to one very imperfect and simple, each grade being useful to its possessor, can be shown to exist . . . then the difficulty of believing that a perfect and complex eye could be formed by natural selection, though insuperable by our imagination, can hardly be considered real." In other words, if we can identify in nature a sufficient variety of advantageous proto-eyes, then it is plausible that one thing could have led to another, and a complex eye could have evolved in a setting of sufficient selection pressure.

Let us rise to the challenge of Darwin's thought experiment, then, with a little help from modern zoology. We are now aware of a rich taxonomy of eye types of varying design and complexity. Nature has provided living creatures with rudimentary light-sensitive patches, fine-tuned biological optics, and just about everything in between. Abundant examples may be found in Yong's *An Immense World*, as well as in *Animal Eyes*, a fascinating scientific treatise by Michael F. Land and Dan-Eric Nilsson.[5]

In 1994, Nilsson and his colleague Susanne Pelger published a theoretical model of eye evolution that goes a long way toward satisfying Darwin's plausibility criteria for the stepwise evolution of eyes.[6] The scientists began with just a flat patch of light-sensitive tissue, an easy enough structure for nature to engineer.[7] They then subjected this simple patch to a series of tiny sequential changes, each producing incremental improvements in sensitivity and/or spatial discrimination. Such a sequence of small steps, they showed, could yield a full-fledged eye, complete with a spherical lens, in less than half a million years. "As far as evolution goes," Yong observes, "that's a blink of an eye."

A lens to collect light and a light-sensitive surface to receive it are two of the key building blocks for modern eyes. To build things up further from there, nature appears to have used two distinct design philosophies. One was to replicate the basic unit of lens and receptor surface many times over, yielding an array of adjacent units pointing in slightly different directions, together composing a compound eye. This brings us to trilobites. To aid them in trawling the sea floor for nutrients, trilobites evolved a compound eye, the structures of which have been preserved well enough in fossilized remnants to offer valuable insight into nature's early efforts to evolve vision. Witness, for example, the trilobite eye shown at the top left of figure 1.1. Each little round bump on the faceted surface is a cylindrical apparatus for focusing and detecting light. Together, the many facets of the trilobite's compound eye divided incoming light into many distinct parcels arrayed in a grid, each element of the grid representing a corresponding portion of the outside world from where the light had originated. The assemblage of views provided by all the distinct facets constituted, by more or less direct correspondence, the image of a scene.

Compound eyes exist in various forms today. They are common, for example, in insects like butterflies. The image at the bottom left of figure 1.1 shows a rendition of how a great spangled fritillary butterfly might see another great spangled fritillary butterfly. To the observing butterfly, which sports a somewhat more evolved set of compound eyes than those used by trilobites, its companion appears as an array of adjacent

**FIGURE 1.1** Compound eye and camera eye. (*Top left*) A microscope image of a trilobite compound eye, circa 500,000,000 BCE. (*Top right*) A close-up photograph of a human camera eye, circa 2000 CE. (*Bottom left*) How a butterfly might see another butterfly through its compound eyes. (*Bottom right*) How we might see a butterfly through our camera eyes.

*Sources*: (*Top left*) E. N. K. Clarkson, "The Evolution of the Eye in Trilobites," *Fossils and Strata* 4 (1975): 7–31, plate 1; (*top right*) air009/Shutterstock; (*bottom left*) Generated by the author using MATLAB (The Mathworks Inc.), with inspiration from *The Orange and the Black*, a series of images created by Henry S. Horn of Princeton University; (*bottom right*) Arthur E. Gurmankin/ Shutterstock.

dots of varying shape and intensity, each dot originating in a single facet of a compound eye. Figure 1.1 clearly illustrates the direct correspondence between the resulting array of dots and the scene they represent. Yes, folks, we have got ourselves a map of spatially organized information— that is, an image.

In the compound eye design philosophy, lenses were primarily used to collect light and thereby to enhance sensitivity. Spatial discrimination was accomplished largely through the orientation of the eye facets, which collect light coming in from various directions. A second design approach relied more heavily on another convenient property of lenses, namely, their ability to focus light. This approach yielded the so-called camera eye, which is the basis of our human visual organs. In a camera eye, like the one shown at the top right of figure 1.1, incoming light passing through a single large lens converges at a focal point, then splays back out to be projected, in inverted miniature, onto a single extended light-sensitive patch containing multiple adjacent light-detecting elements. Eliminating the requirement for multiple lenses frees up real estate to pack in lots of detectors (such as the more than one hundred million rod and cone cells in our retinas). This yields high-resolution images like the picture of the great spangled fritillary butterfly shown at the bottom right of figure 1.1.

Figure 1.2 illustrates the internal structure of a human camera eye and shows how such an eye captures an image. This mechanism of image capture would eventually be replicated in the technology of—you guessed it—cameras.

What about the other features of human eyes that I mentioned earlier when I asked you to look in a mirror? Consider color vision, for example. What we perceive as distinct colors correspond to distinct light frequencies or distinct combinations of light of various frequencies. Lower frequencies lie closer to the red end of the color spectrum and higher frequencies closer to the blue end. Discrimination among frequencies is accomplished

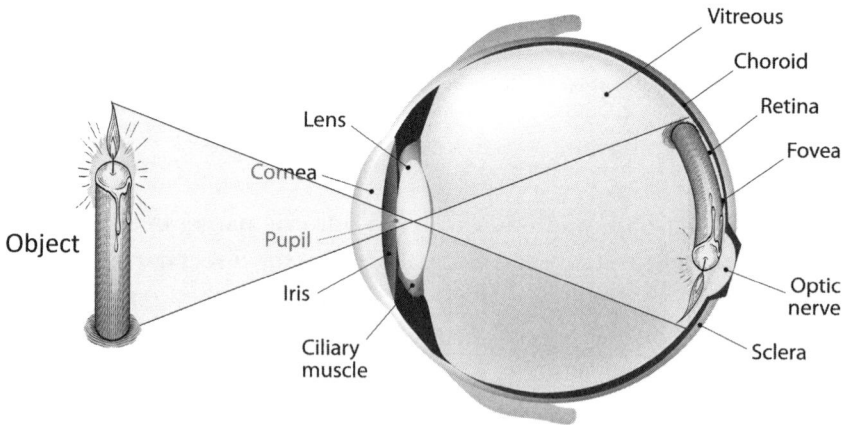

**FIGURE 1.2** Inside the human eye.

*Sources*: (*Candle*) Created by the author using DALL-E 3 (OpenAI's image generator, in the ChatGPT 4o portfolio); *(eye diagram)* Designua/Shutterstock.

by shape-shifting molecules called opsins, which convert light into electrical signals that our brains can detect. Light arrives, and the nerves behind our eyes light up with electrical activity, which can then be carried anywhere it needs to go. Interestingly, across species, opsins appear to have a single common evolutionary ancestor, even though eyes are believed to have evolved in many distinct ways from diverse origins. The shape-shifting portions of opsin molecules are called chromophores (literally, "bearers of color"). They come in many varieties with differing sensitivities to different frequencies of light. In particular, the cone cells in our eyes have three distinct opsin chromophores, which are preferentially sensitive to frequencies we associate with red, green, and blue. The fact that we can see many subtle gradations in color, rather than just single shades of red, green, and blue, or simple mixtures of those shades, is the result of clever evolutionary engineering and intelligent postprocessing. The frequency sensitivities of our various photoreceptors, while peaking in different places, actually overlap so that any given light frequency elicits a response in more than one chromophore. The balance of signal strength in photoreceptors carrying the three types of chromophores results in our perception of color. A signal received more strongly in "red" cones than in "green" or "blue" cones, for example, causes us to perceive a redder hue. Networks of neurons in our brains also perform more advanced comparisons, working the various signals against one another to distinguish subtler shades. The overlapping multidetector arrangement in our eyes is terrifically efficient, enabling the discrimination of a large set of frequencies with a small number of detector types. Humans would eventually use similar tricks to create color photography, which captures rich palettes of color using a small number of light-sensitive elements. When the science of chemistry became sufficiently advanced, humans would even learn how to design their own light-sensitive opsin-like molecules. With sufficient engineering under our belt, we would also build electronic detectors with custom-designed frequency sensitivities. The result would be a modern revolution in imaging that would teach us even more about our own biology.

So much for color. What about depth? Our camera eye arrangement allows us to tell left from right and up from down. But how do we tell how near or far an object may be from us? Clearly, this is a crucial function if one is skirting the edge of a cliff or if one wishes to avoid occupying the same space as a pouncing predator. Our brains use an assortment of methods to deduce depth, one of which may explain why we have two eyes rather than one. It turns out that the color-vision trick of multiple receptors with differential sensitivity was so good that it bore repeating. From our two separate eyes, our brains receive two distinct images of any given scene. These two images are obtained from slightly different angles and therefore have slightly different appearances. By comparing the two images, our brains can triangulate depth. This process of stereopsis, which converts raw visual data into new spatially resolved information, serves as a nice model for various forms of modern image reconstruction, in which digital computers take the place of biological brains.

One other enviable design feature of our eyes is their adaptability. Through the ministration of small, finely controlled muscles, we can shift our field of view without turning our head, and we can even adjust our focus at will. Each of our eyes has an iris that can dilate and contract to control the amount of light it lets in. Our eyes are infused with a continuously refreshed transparent fluid (with the amusingly formal name of vitreous humor) to maintain a consistent shape without blocking precious light. We have hair-trigger blink reflexes to prevent things other than light from getting into our eyes, and our eyelashes are also on guard duty keeping foreign bodies from gumming up the sensitive works. Countless analogues of moving parts like these can be found in human imaging technology.

If, in focusing on how our eyes work, I have managed to portray human eyes as the singular pinnacle of evolutionary development, then let me set the record straight right now. Clearly, human eyes are impressive feats of evolutionary engineering, but let's not be too pleased with ourselves. Despite their impressive capabilities, human eyes are not the best eyes the world has seen. This is true for nearly any criterion of quality one might choose. For proof, look no further, once again, than Yong's *An Immense World*, which describes a dizzying variety of eye types and visual strategies. Yong makes the case that our sensory window onto the world is just a tiny sliver of the entirety of what can be sensed by other forms of life. Some animals use scent or sound to see, and others orient themselves using probes that are naturally invisible to us, like electric and magnetic fields (which we humans only learned to detect with manufactured tools comparatively late in our technological development). In the narrow realm of light-detecting eyes, birds of prey have far greater visual acuity than the most clear-sighted human. Cows can see in almost all directions at once. Many animals can see into the ultraviolet, and some can distinguish different polarizations of light, which are invisible to our naked eyes. The mantis shrimp has twelve distinct chromophores as opposed to our comparatively paltry three.[8]

Some species of scallops have far more than two eyes. Sea urchins and brittle stars appear to have anywhere from hundreds to thousands of eyes (or at least photoreceptors) covering most of their body surface.

In fact, human eyes represent something of a compromise among various design criteria. Observations from nature suggest that the type of eye one finds in any particular organism generally depends sensitively on the tasks for which it evolved. For example, some of the eyes sported by certain species of jellyfish are weighted so that they always point upward, giving the animals a good view of the mangrove forests above them. Earthbound predators like humans often have forward-facing eyes so that they can focus on their prey without bothering to watch their back, whereas prey animals tend to have a wider view so that they can catch sight of potential threats before those threats are close enough to pounce. Airborne predators, on the other hand, may be blind to what is right in front of them since they tend to approach their prey from above. The giant squid appears to have evolved exquisitely sensitive eyes the size of soccer balls in order to detect the faint blush of bioluminescence that its principal predator—the sperm whale— elicits from surrounding creatures when it moves through the deep ocean. There is actually evidence that some creatures once had powerful visual capabilities that they eventually lost over time as they adapted to visually undemanding environments. The key to eyes, it seems, is not their imaging prowess per se, but rather the specific information they convey.

Let us close, then, by considering the information that eyes have to offer. This all comes down to the properties of light. Light is a remarkably powerful probe of the natural world. It is in many ways the perfect spy. First of all, light is plentiful in our corner of the universe. We have a nearby star—our sun—whose thermonuclear emissions envelop our planet and all its creatures in light, like a second sea. Thus, it is only natural that early sea creatures should have evolved mechanisms to detect that light. Meanwhile, most materials on our planet or elsewhere can be induced to produce light when hot or suitably bothered. This means that there are rich sources of light everywhere, whether at modest local scales—in the form of fire, lightning, or glowing magma—or else at the scale of galaxies. Second, light is fast. As Einstein taught us, in fact, there is precisely nothing faster. If unobstructed, light can travel a distance equivalent to seven circuits around the world in a single second.[9] This means that the information it provides is prompt and actionable. When opaque objects do get in its way, light is absorbed, reflected, refracted, or scattered, with the color, intensity, and direction of the resulting light beams speaking volumes about the shape and composition of the objects in question. So, light carries with it a detailed record of all the many and varied things it encounters as it goes along its merry, speedy way. In summary, when it comes to selecting a probe that offers rich spatial information about the world, it makes perfect sense that evolution would have gone to so much trouble to see the light.

Why have I gone to so much trouble to describe the evolution and operation of eyes in a book about human-made imaging methods? I have done so because natural eyes are a compelling model for many current imaging methods and technologies. Nearly all of the visual features discussed in this chapter may be found, either replicated directly or cleverly adapted, somewhere in the history of imaging. Eyes take advantage of many of the physical properties of light, and they also give us ample hints about how to make good use of probes other than light. They show us how we see, and they suggest to us how we might go about inventing new ways of seeing.

As you have seen in this chapter, the biology of vision is variegated. The physics, however, is simple. We need sensitivity, and we need spatial discrimination. We collect light, and we bend it to our will. These fundamental building blocks of vision are the first things that people learned to copy.

After we had eyes, and brains, it took us a fair bit of trial and error to emulate nature and improve upon it. Humans spent many millennia getting practice with making images of our own, from primitive cave paintings to imperial mosaics to grand painterly feats of eye–hand coordination. We sussed out the principles of geometry—how objects are organized in space—and we worked these principles into our images to render them more realistic. We experimented with various natural materials, noting curious properties that might one day come in handy. Eventually, we had everything we needed to create our own imaging revolutions.

The next two chapters expose two remarkable tricks of perspective that ultimately transformed our ability to see. The first transformation, enabled by incremental improvements in technology over the course of centuries, emulates the built-in focusing function of our eyes in order to extend their capabilities, allowing us to see farther and more clearly than we ever could before. The second was instead the result of a sudden and stunning discovery. It did something our eyes were never built to accomplish, ultimately giving us x-ray vision.

# 2

## Augmenting Nature

How the ability to bend light gave us the gift of magnification.

I t was a small thing, really. Some distant ancestor of ours picked up a clear piece of rock crystal and looked through it. What she saw changed the course of history.

In chapter 1, we learned that all the mechanisms required for advanced imaging technology were lying in wait for eons—hidden, if you will, in plain sight. The earliest humans already carried elaborately engineered imaging devices with them, and within them, wherever they went. But it would be a long time until humans had either the tools or the stomach to dissect out the inner workings of their own eyes.

In the meantime, they had to work with materials in the world around them. Using some of those materials, our ancestors learned how to do a simple but monumental thing. They learned how to bend light. And this little twist would ultimately bring the cosmos within their reach.

That's it. Bending light.

There are entire libraries devoted to the tools of astronomy. There are countless viewing devices on display in museums across the globe or perched on tripods in the bedrooms of children and the attics of hobbyists everywhere. There are observatories the size of mountaintops, in which questing scientists reach out across vast spans of space and time. But it all started with bending light. Why? Because bent light gave us the power to magnify.

Before optics, we had our eyes and our imaginations. Don't get me wrong; they brought us far. When we mastered the art of magnification, though, we tumbled down the rabbit hole to a strange new wonderland. What was small became large, and what was unimaginably far became tantalizingly near. With light-bending devices, we were able to see far out into the universe. We could also witness firsthand the microcosm hidden in a drop of pond water. Our sense of scale shifted, and the vast "outer space" that had been intuited by many of the great ancient human civilizations was suddenly brought down to Earth. We also became familiar with the concept of "inner space," where tiny organisms invisible to the naked eye went about their mysterious business.

How does bent light result in magnification, you may ask? Well, that is our story in this chapter. It takes us on a path from the camera obscura to the modern camera, with stops along the way for telescopes, microscopes, and other marvels. Come along, won't you?

Some of the earliest human forays into manipulating light began by simply blocking it. We all know from early childhood that you can create shadow pictures on a wall by holding your hand in front of a light source. One must imagine that early cave-dwellers were similarly aware of this phenomenon. They may also have been aware of another phenomenon that produces what you might say is the opposite of a shadow picture. When a small hole opens into a dark space, the light coming through that hole can actually make a light picture of things on the outside. This configuration—a pinhole that admits only a small quantity of light into a dark room—came to be called a camera obscura (which, unimaginatively enough, means "dark room" in Latin).

Figure 2.1 shows how a camera obscura works. Light from the outside world travels through the pinhole and creates an image on the opposite wall. Since light coming from any particular part of an object or a scene will make it through the hole only if it is traveling at just the right angle, it will show up at just one spot on the wall, and the image will be a faithful (if inverted) representation of the object or scene.[1] This basic geometry should remind you of what happens in the human eye. That is why we call it a camera eye.

Plato's famous allegorical cave, with suggestive images projected onto the walls, was something tantalizingly close to a camera obscura.[2] Some argue that images inadvertently projected through pinholes in tents or screens may even have inspired primitive paintings on the walls of actual

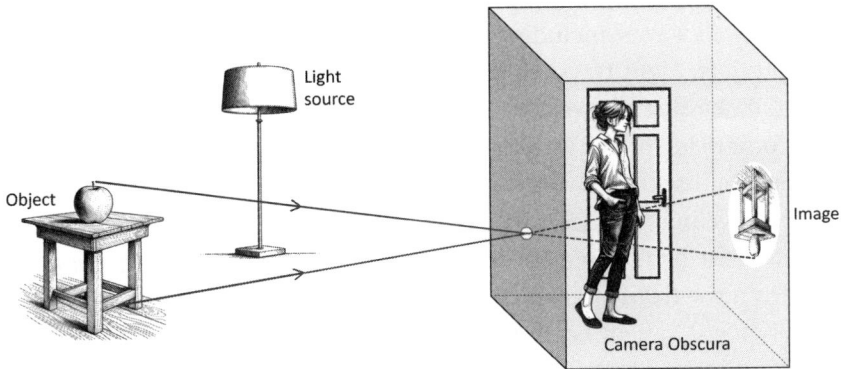

**FIGURE 2.1** A camera obscura.

*Source*: Created by the author with DALL-E 3 (OpenAI) and Microsoft PowerPoint.

caves. It is indisputable that camera obscuras were used to observe eclipses and project images of people from as early as 500 BCE (as recorded in early writings from Han China). By the fifteenth century CE, they had been used to copy scenes, study the behavior of light, make astronomical observations, and provide entertainment at parties. Do you know who was a particular fan of the camera obscura? Leonardo da Vinci, the famous polymath who rendered the original *Mona Lisa* for our hungry eyes. He studied camera obscuras in detail, eventually reaching a very modern conclusion that the human eye operated on similar principles. Long before Leonardo, the principles of the camera obscura were laid out by the influential mathematician and astronomer Ibn al-Haytham in his *Book of Optics*, a seven-volume treatise written over a decade in the early eleventh century.

One key structure in the human eye that is missing from a simple camera obscura is a lens. The word *lens* derives from the Latin for "lentil," whose biconvex shape, curving outward on both sides, is characteristic of many lenses. One of the first known artificial lenses is the so-called Nimrud lens—a three-thousand-year-old piece of roughly ground rock crystal discovered more than a century and a half ago in the ruins of the Assyrian city of Nimrud, in modern-day Iraq. Scholars believe it may have been used as a "burning glass" to concentrate sunlight. Or it may just have been meant for decoration. Regardless, it is known that lenses were objects of curiosity and active exploration dating back to ancient Mesopotamia, Egypt, and Crete.

How does a lens work? It relies on the phenomenon of refraction experienced by all waves, including light waves, at any interface between different materials. Light travels just a bit slower inside the lens than it does in air, and if one side of the wave hits the lens first, it slows down with respect to the other side of the wave until that other side also enters the lens. The net result is that the light wave is tilted away from the direction it was previously traveling. For a flat piece of crystal, the wave would be tilted back as soon as the light exited back into air. Since the far edge of the lens curves away from the front edge, however, the tilt just increases. As shown at the top of figure 2.2, light heading directly toward a lens tilts more the farther it gets from the center line. This means that parallel light rays are bent together toward a focal point on the other side of the lens.[3] In a burning glass, the focal point is a literal hot spot where incoming light converges.

Notice that the focal point of the lens behaves in many ways like the pinhole in a camera obscura. Light focused inward by the lens spreads back out beyond the focal point. Also like a camera obscura, therefore, a lens can produce an image. One key difference is that a lens is generally bigger than a pinhole, so light from the same point on an object can travel different paths through a lens. This means that more light passes through a lens than through a pinhole, and the resulting image is brighter. It also means that the image is only in focus at a particular distance from the focal point, where light coming from any given point on the object converges back to the same point in the image.[4] This is illustrated at the bottom of figure 2.2.

Eventually, people began to insert lenses into the widened openings of camera obscuras. They realized that brighter images were worth the extra engineering effort of matching lens shape to room size in order to keep images in focus. Nature, of course, had come to the same conclusion long before. The lenses in our eyes concentrate light and increase our visual sensitivity. The compromise, to avoid blurring and therefore preserve spatial discrimination, is that the retina must be located at just the right distance from the lens. This is the origin of nearsightedness and farsightedness—conditions that occur when the shape of someone's eyes is not quite matched to the focusing range of their native lenses. As early as the thirteenth century, clever glassmakers had figured out how to clear up congenitally blurry vision by placing corrective lenses in front of the eyes. For the first time in history, humans had begun to make routine modifications to their natural vision.

Lenses can do more, though, than just clear things up a little. Figure 2.3 illustrates the operation of a magnifying glass. A magnifying glass is just a lens placed close to an object of inspection, which, when viewed from

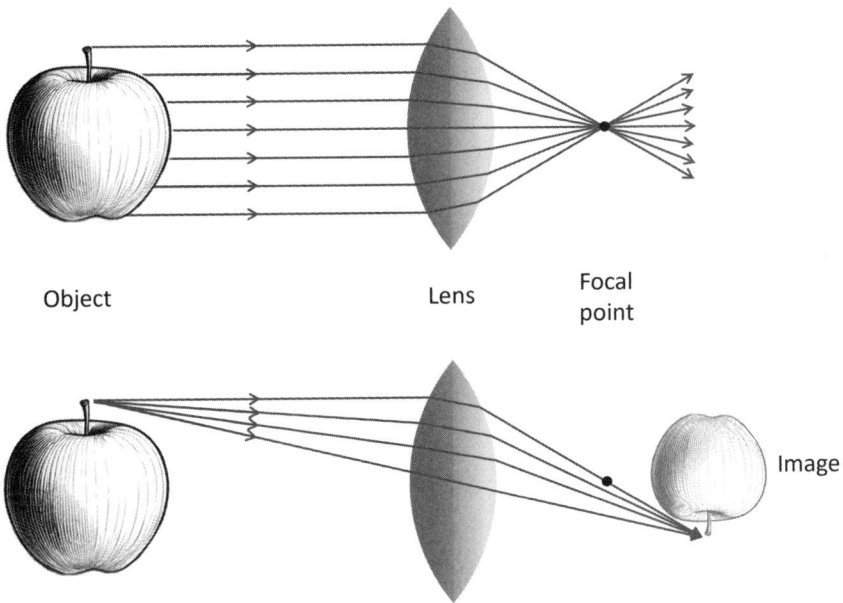

**FIGURE 2.2** A lens bends light.

*Source*: Created by the author with DALL-E 3 (OpenAI) and Microsoft PowerPoint.

the other side of the lens, appears larger than it truly is. The figure shows how this appearance is a direct consequence of the bending of light. Light emerging from each point on the object is bent in such a way that it appears to be coming from a corresponding point on a much bigger object. The visual centers in our brains are trained to trace light back to where it came from, and when they do this with the bent light passing through a magnifying glass, they arrive at the conclusion that we are seeing something larger than life, complete with all the details that might otherwise have escaped our notice. The resulting optical illusion is not merely a source of fascination for children—it has real and lasting power. Magnifying glasses may be used to inspect objects in fine detail when our unaided eyes are not enough. Tools like jeweler's loupes and detective's hand glasses do not merely replicate the appearance of an object. They transform it.

While the distorting power of lenses was known from early on, the extent of the resulting transformations remained modest for millennia. There are practical limits to how much magnification one can achieve with any given lens, and simple magnifying glasses don't work at all with faraway things.

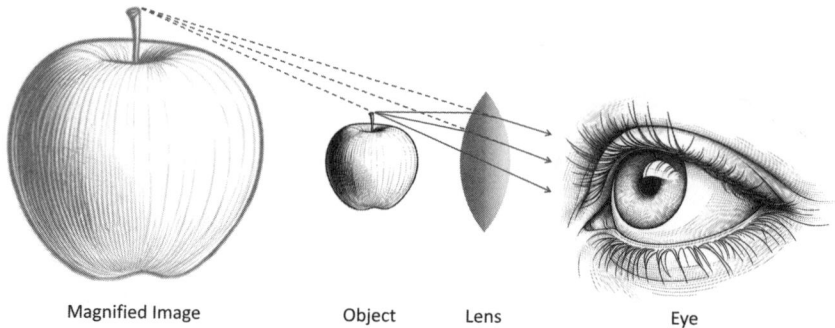

Magnified Image              Object          Lens                     Eye

**FIGURE 2.3** How bent light can result in magnification.

*Source*: Created by the author with DALL-E 3 (OpenAI) and Microsoft PowerPoint.

Early in the seventeenth century, though, ingenious tinkerers figured out how to get around all these limitations in one fell swoop. Their solution? Multiple lenses.

By the end of the sixteenth century, eyeglasses were all the rage, and there is evidence that they were being manufactured in many European countries. This meant that there were craftsmen across the continent who were paid to spend their days fiddling with lenses. It was actually a Dutch spectacle-maker named Hans Lippershey who had the first reliably documented idea for a telescope with two lenses and two stages of magnification. On October 2, 1608, Lippershey filed for a patent on an instrument "by means of which all things at a very large distance can be seen as if they were nearby." His application was denied! But the time was clearly ripe for this invention. In fact, the rejection of Lippershey's claim may well have been due to the existence of similar claims by other spectacle-makers at the time. Jacob Metius, also from the Netherlands, applied for a patent weeks later. The contention of Johannes Zachariassen that his father, Zacharias Janssen, had invented the telescope (and, for good measure, the microscope) nearly twenty years earlier would be repeated in history books for centuries to come.[5] This would not be the last time that controversy attended key innovations in imaging.

Indeed, the controversies were just getting started. Galileo Galilei heard about the Dutch "perspective glass" in Venice in June of 1609, and, as he told the story, designed a new telescope the very next day. (Galileo's was the first instrument to be called a telescope.) Before long, Galileo turned his enhanced eyes to the sky, with famous results. He discovered satellites of Jupiter, surface features on the moon, and sunspots, among other surprises. These observations brought Galileo both fame and heartache. The story of his treatment at the hands of the religious authorities of the time, who sought to keep the heavens innocent of human meddling, has been told many times. I will not repeat it here, other than to note that the Galilean controversy did not come to an end until 1992, when Galileo was officially cleared of wrongdoing by the Vatican.

How did these early telescopes work? Figure 2.4 illustrates the essential concepts. The trick was to use one lens (now known as an objective lens) to produce a first image (called a primary image) of a faraway object, just as one would do with a camera obscura. A second lens (an eyepiece lens) was then trained on the primary image to magnify it. The advantage of this arrangement was that a lens designer could compute exactly where the primary image would show up and could therefore control the magnification of the final image by careful placement of the eyepiece lens.[6]

The bottom illustration of figure 2.4 shows that, with a slight change in lens configuration, the principle of two-stage magnification also works to produce substantial enlargements of small, nearby objects. This fact was not

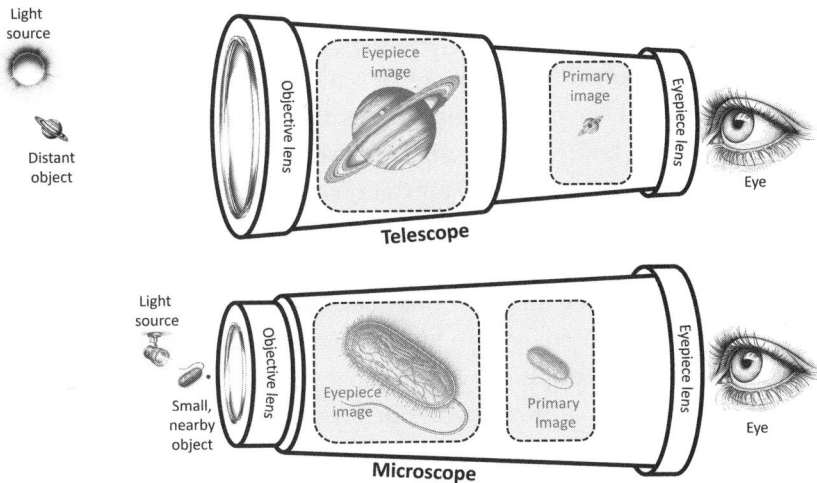

**FIGURE 2.4** Combinations of lenses produced humankind's first powerful imaging tubes: telescopes and microscopes.

*Source*: Created by the author with DALL-E 3 (OpenAI) and Microsoft PowerPoint.

lost on those early-seventeenth-century glassmakers. Microscopes have an origin story similar to that of telescopes—complete with some of the same controversies. Lippershey, Janssen, and Galileo were all in the fray. That said, while the perspective afforded by telescopes quickly inflamed both curiosity and conservative ire, it took a little longer for people to appreciate that microscopes also had revolutionary revelations to offer. Microscopes were viewed largely as a novelty until naturalists in Britain and on the Continent began using them in earnest in the second half of the seventeenth century. Robert Hooke, sometimes called "England's Leonardo,"[7] recorded his observations through microscopes (and telescopes) in his 1665 book, *Micrographia*. He penned the earliest known direct observation of a microorganism (a fungus called *Mucor*) and coined the term *cell* based on the tiny structures he saw in plant samples. Antonie van Leeuwenhoek was a Dutch contemporary of Hooke's who was similarly drawn to microscopes. His observations of bacteria helped to establish the field of microbiology.

As the examples of Galileo, Hooke, and others illustrate (see figure 2.5), even comparatively simple tools of enhanced vision can lead to startling discoveries. Humanity's first close encounters with extraterrestrial moons and earthbound microbes may be traced directly to the placement of a couple of lenses inside a tube. The great twentieth-century physicist Freeman Dyson once wrote that "The great advances in science usually result from new tools rather than from new doctrines. . . . Every time we introduce a new tool, it always leads to new and unexpected discoveries, because Nature's imagination is richer than ours."[8] History has shown that few tools are more powerful for making discoveries than imaging devices. What is discovery, after all, but the uncovering of what was once hidden? While eyes may have begun as a survival tool for Cambrian creatures, they evolved into key drivers of curiosity about the world. Likewise, as soon as we humans were able to improve upon natural eyes, we launched a virtuous cycle of invention and discovery that has continued to this day. New imaging tools have generated new discoveries, which have motivated and provided building blocks for the development of still more powerful tools. Seeing begets understanding, which in turn enables better ways of seeing.

The years, decades, and centuries that followed the invention of telescopes and microscopes have seen an explosion of activity in optics—the science of manipulating light. No lesser figures than Isaac Newton and Johann Wolfgang von Goethe dove deep into the theory of optics, and brilliant practical innovations abounded across the globe. Curved mirrors

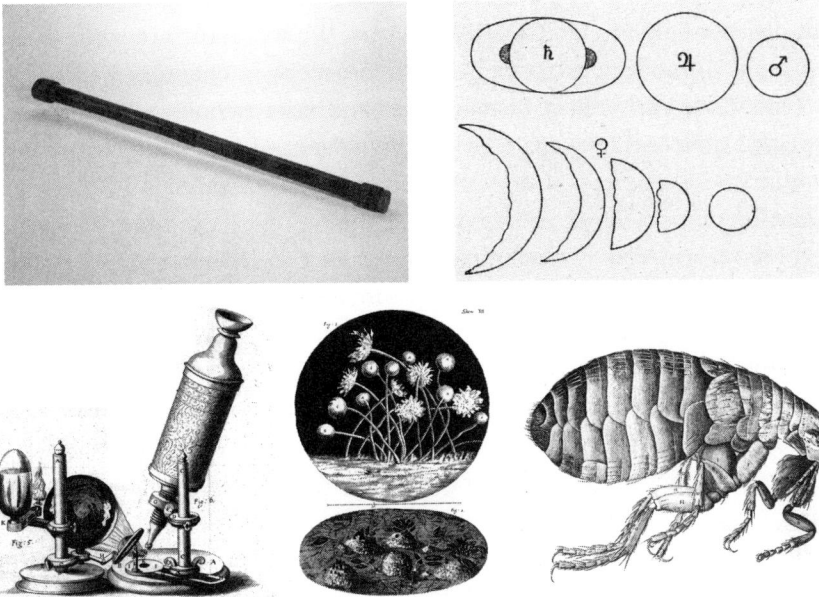

**FIGURE 2.5** Revolutionary discoveries made with telescopes and microscopes. (*Top*) Galileo's original telescope (c. 1609) and sketches of what he saw. (*Bottom*) Hooke's microscope and drawings from *Micrographia* (1665).

*Sources*: (*Top left*) © Museo Galileo, Florence, Italy; (*top right*) Public domain via Istituto di Linguistica Computazionale (https://solarsystem.nasa.gov/resources/482/galileos-phases-of-venus-and-other -planets/); (*bottom*) Public domain via the Wellcome Collection.

were introduced into the light-bending mix to get around some annoying imperfections associated with lenses. In the realm of astronomy, light-collecting mirrors grew in size, improving both sensitivity and spatial discrimination until high-powered telescopes were larger than houses. As technology advanced, devices were constructed to capture frequencies of light other than the visible (radio waves, infrared, ultraviolet, and beyond). Telescopes girdled the earth, and some were launched into orbit. At this very moment, distant cousins of Lippershey's or Janssen's or Galileo's tube are hurtling through the void of space that telescopes first helped humankind to discover. Meanwhile, the modern imaging armamentarium contains all kinds of powerful microscopes that have revolutionized our understanding of inert materials and biological tissues alike. In 2020, it was electron microscopes—particularly high-powered devices that use streams of charged particles in place of light beams—that first showed us the tiny but distressingly mighty COVID-19 virus. We will return to some of these triumphs of modern astronomy and microscopy in chapters 5 and 6.

One other early effect of telescopes and microscopes, which would be repeated time and time again as imaging advanced, was a fundamental shift in humankind's perspective on the universe and our place in it. The Copernican Revolution, which is often used as the canonical example of a scientific revolution, was also a revolution in perspective. It began in the minds of deep thinkers like Copernicus but was anchored in the end by observations through Galileo's telescope. After the Copernican Revolution, it was increasingly difficult for humans to see ourselves as occupying the center of the universe. Instead, we understood ourselves to be passengers on a rock hurtling through space. At the same time, thanks to microscopes, people began to realize that inner space was also brimming with unexpected passengers. This contributed to dizzying shifts in humanity's sense of scale. By 1726, *Gulliver's Travels* could gleefully depict the selfsame traveler as either gargantuan or miniscule, depending upon whether he was standing next to a tiny occupant of the island of Lilliput or a giant from the land of Brobdingnag.

The source of all of these many and varied influences may be summarized in a strikingly simple recipe: (1) Bend light. (2) Repeat. This straightforward transformation of our accustomed lines of sight catapulted us beyond the limits of natural vision. It showed us that an image can be more than just a copy of the visible world. It also set the stage for ever more abstract transformations that would constitute the subsequent history of human imaging.

Before we get to the next key transformation, though, we have one last stop on our whirlwind tour of optics. The inventors I have paraded before you so far were masters of magnification, but they still had a challenge when it came to recording the beautiful images brought to them by telescopes and microscopes. The time-honored way of recording images was, of course, to make painstaking copies by hand. But what if you were observing a once-in-a-lifetime astronomical event or witnessing a dynamic interaction among microorganisms in real time? Like Galileo or Hooke, you would have had to draw what you had seen. And what if you were not a master draftsman endowed with a vivid memory? You would certainly have wished for a way to capture an image automatically in the moment. Enter the modern camera and its secret weapon: film.

Curious minds had long been searching for mechanical means of recording visual scenes, but such means had proven elusive. Artists and hobbyists were left to trace the "light paintings" in camera obscuras line by line. With lenses and mirrors, one could bend light all one liked, and one could construct elaborate frames to position the optical elements just so, but ever-transient light beams simply refused to be frozen in place on demand. It took chemistry to do the job.

Just as chemical chromophores are the secret to light sensitivity in our eyes, light-sensitive materials were the origin of photographic film. It had been observed since at least the thirteenth century that certain silver salts darkened when exposed to sunlight. Early in the nineteenth century, the serial inventor Joseph Nicéphore Niépce invented a process he called heliography, meaning "sun drawing." Inspired by the emerging art of lithography, he coated sheets of stone, metal, or glass with bitumen, a substance that hardens on exposure to light. He then used a solvent to rinse away the soft, unexposed bitumen and etched the surfaces not protected by bitumen with acid, leaving grooves wherever light had not shone. Using his light lithography procedure, Niépce created the first recorded permanent photographic image in 1822.[9] He spent the next seven years experimenting with and improving on heliography.

In 1829, Niépce entered into a partnership with a larger-than-life character named Louis-Jacques-Mandé Daguerre. Daguerre was a panorama painter, a theater designer, and a master of dramatic illusion. The partnership brought fame to Daguerre, though it ended in heartache for the family Niépce, who would claim that Daguerre had passed off Joseph's ideas as his own. After Niépce died suddenly of a stroke in 1833, Daguerre continued to experiment and ultimately developed a process using sheets of silver-plated copper tarnished by controlled exposure to light. He modestly named this process after himself. Figure 2.6 shows equipment used for and selected early outcomes of the daguerreotype process. When Daguerre first saw the fruits of his labors in the form of a developed image, he is said to have proclaimed, "I have seized the light—I have arrested its flight!"[10]

Thanks in part to Daguerre's flair for showmanship, the daguerreotype succeeded in capturing the popular imagination. Famous images that no human hand had traced could now be passed around for comment by anyone who could get their hands on them. Photography would go on to take the world by storm. Color film (emulating the eye's three-color trick) would add newly realistic zest to images. Daguerreotypical snapshots would be upgraded to moving pictures: rapid series of images captured over time and played back in a realistic temporal sequence. Hollywood, California would be converted from an agricultural enclave to a mecca

FIGURE 2.6 Completing the journey from camera obscura to camera. (*Left*) A daguerreotype camera, 1839. (*Center*) An 1837 daguerreotype image by Louis Daguerre. (*Right*) An 1844 daguerreotype image of Louis Daguerre.

*Source*: Wikimedia Commons

of image generation. Whereas telescopes would grow larger, photographic devices would get smaller and smaller. And digital means of capturing images would come to replace film.

Even in the nineteenth century, though, film had already sown the seeds of digitization. It would eventually come to be understood (thanks to the resolving power of microscopes) that light-sensitive film is made up of small grains, each of which can change color on exposure to light. This might remind you of the discrete photoreceptor cells in our camera eyes or the individual facets of a butterfly's compound eye. Indeed, if lenses are substitutes for the eye's built-in focusing apparatus, then film can be seen as a stand-in for our retina. Whereas telescopes and microscopes enhance the eye, film replaces it.

How did Daguerre engineer the controlled light exposure needed to tarnish his film? With a portable camera obscura: a box with a hole in the front. (Over time, the "obscura" in the name faded into obscurity, leaving us with just the familiar "camera.") In a way, then, we have found our way right back to where we started our journey at the beginning of this chapter. We began with a dark enclosure that let a little light in, presenting ghostly images to our eyes. We have ended with a dark enclosure that could take the place of eyes, using lenses for light collection and magnification as needed, a shutter (like our iris) to let in the light, and a sheet of film at the back to record the resulting images for posterity.

# 3

# Seeing It Through

How the discovery of a mysterious penetrating probe allowed us
to see through previously opaque things.

n November of 1895, Wilhelm Conrad Röntgen accidentally captured
something new on film. As he worked feverishly in his laboratory to
characterize the properties of this mysterious something, he quickly
realized that it could travel through tissues of the body, leaving a shadowy
record of what it had encountered along the way. He asked his wife, Anna,
to place her hand on a cartridge of film and created a skeletal image that
shook the world. So began humanity's intimate association with x-rays.

In 1895, Röntgen was a professor of physics at the University of Würzburg
in Germany. If you had visited his laboratory back then, you would have
encountered glass tubes of various shapes and sizes arrayed on benches
beneath a rectangular wall-mounted pendulum clock to mark the time
(figure 3.1). You would have seen cardboard screens coated with fluorescent material that glowed upon exposure to cathode rays—a topic of study
for Röntgen and a common fascination of physicists at the time. You would
also have seen assorted photographic equipment since Röntgen was an
avid photographer of family vacations and scientific experiments alike.

On November 8, Röntgen was in the lab experimenting with a so-called
Crookes tube, a device designed to generate cathode rays. It was known

35

**FIGURE 3.1** Wilhelm Conrad Röntgen (*left*), Röntgen's laboratory (*center*), and Anna Bertha Ludwig (*right*).

*Sources*: (*Left*) Nicola Perscheid, Hofphot., Berlin W. 9/ETH Zürich, CC BY-SA 3.0, via Wikimedia Commons; (*center*) public domain via Wikimedia Commons; (*right*) public domain via Google Arts & Culture.

that cathode rays (which would later come to be identified as streams of electrons) could not travel far from the tubes that generated them. So Röntgen thought he was safe in placing a spare fluorescent screen on a chair a few feet away. When he switched on the current to power the tube, though, he noticed a soft glow emanating from the screen.

Röntgen's genius lay in not ignoring that glow. Others had not been so attentive. Back in 1890, a photographer visiting the lab of American physicist Arthur Goodspeed had left two coins atop a pile of photographic plates during a cathode ray demonstration, and had subsequently found circular shadow images on the plates when he developed them. Not knowing what to do with these mystery smudges, however, he set them aside—until he heard the news of Röntgen's discovery years later.

Röntgen realized that cathode rays could not have traveled far enough from their tube of origin to reach his glowing screen. Being an avowed experimentalist, committed to exploring observations without prejudice or theoretical predisposition, he began a new round of experiments. He worked furiously in isolation for weeks, carefully cataloging the conditions under which he could reproduce the glow and also exploring means of blocking it. He found that whatever was producing the glow on the screen could travel an appreciable distance and could be stopped over shorter distances only by very dense materials such as lead. Small samples of dense material would leave a shadow on the screen.

At some point, Röntgen placed his hand between a Crookes tube and a phosphorescent screen, becoming the first human to see the shadows of his own bones beneath his intact skin. He recorded his experiments with photos, and he also found that the mystery rays causing his screens to glow could be captured directly on photographic film. On December 22, he invited his somewhat bemused wife, Anna Bertha Ludwig, into the laboratory and convinced her to lay her hand on a photographic plate with the cathode ray tube switched on for fifteen minutes. The resulting spectral picture of her skeletal hand with its prominent wedding ring quickly became one of the most famous images in history (figure 3.2). Upon viewing the image, Frau Röntgen is said to have imagined that she was seeing her own death. As countless others would soon come to know, she was certainly seeing the future.

In the waning days of December, Röntgen's initial report, "On a New Kind of Rays," was published in the *Proceedings of the Physico-Medical Society of Würzburg*. He mailed copies to various leading scientists across Europe and Great Britain, and soon the cat was officially out of the bag. When it came to naming his new rays, Röntgen settled on the simple label "x" for the unknown.

The mystery turned out to be irresistible—to scientists and the lay public alike. In the first chapter of her illuminating treatise on the history and societal impact of medical imaging in the twentieth century, Bettyann Holtzmann Kevles describes what followed Röntgen's discovery as nothing short of "x-ray fever."[1] Kevles identifies forty-nine books and 1,044 papers published on x-rays in 1896 alone. As a professional scientist, I can attest that this is a remarkable outflowing of paper for a singular new discovery, even in today's age of highly connected and citation-hungry scientific research.

The kudos flowed in. Röntgen instantly became a public figure despite abandoning his study of x-rays not long after his seminal discovery. The first-ever Nobel Prize in Physics went to Röntgen in 1901 "in recognition of the extraordinary services he has rendered by the discovery of the remarkable rays subsequently named after him."[2] Arguably, like Einstein after him, Röntgen had become such a household name that he conferred more credibility on the fledgling prize than the prize brought to him. Let it not be forgotten that it was imaging that gave the Nobel Prize its earliest luster.

Many other laureates would follow directly in Röntgen's footsteps. Among them was Henri Becquerel, who received the 1903 Nobel Prize in Physics for discovering spontaneous radioactivity—which would subsequently form the basis of nuclear imaging methods. Becquerel, who made his discovery in March of 1896, is said to have been inspired by hearing

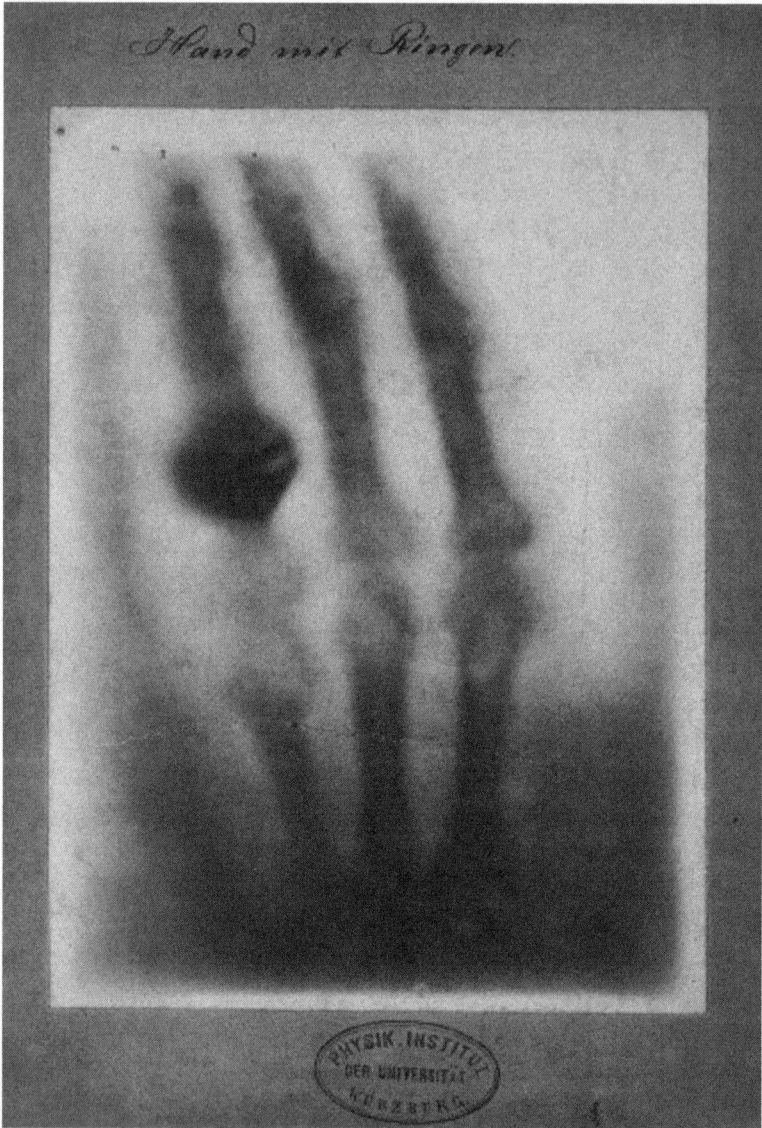

**FIGURE 3.2** Röntgen's x-ray image of Anna Bertha Ludwig's hand, with ring.

*Source*: Wikimedia Commons

Röntgen's letter to Henri Poincaré read aloud at the French Academy of Sciences. Becquerel shared his 1903 Nobel with Pierre and Marie Curie. Marie, who ultimately one-upped Röntgen by receiving two Nobel Prizes, would become an advocate for diagnostic uses of x-rays.

And then there were all the scientists who did their enabling work outside the limelight. X-rays are often presented as a paradigmatic "eureka" discovery—entirely unexpected, noted accidentally by a single person, and with an uncontested pedigree. As I trust you are beginning to see, uncontested discoveries are quite rare in the history of imaging, not to mention in the history of science and human experience more broadly. Many elements set the stage for Röntgen's serendipity, including the development of photographic film, phosphorescent screens, and, of course, the Crookes tube.

Indeed, one of the developers of the modified Crookes tube Röntgen had used—a man by the name of Philipp Lenard—became quite sore at the attention lavished on Röntgen. He had experimented with cathode rays and fluorescent screens and had even published reports of a telltale glow, but he had neglected to record it or investigate its origin. He demanded credit and lambasted Röntgen for not acknowledging him sufficiently. A subsequent Nobel Prize of his own did not appease him, and in accepting the prize in 1905 he made a point of condemning the 1901 decision. Röntgen was nonplussed. To quote Kevles, "He could not know that Lenard's reaction, if extreme, would be emulated many times . . . in the coming years by other image-making entries in the Nobel sweepstakes."[3] I myself have been witness to one such bitter Nobel controversy in my own field of MRI—but I'll get to that in good time.

Leaving aside the infighting of ambitious scientists, the public's fascination with x-rays continued. The new rays blasted their way into news articles, commentaries, and cartoons. They suffused popular prose and poetry alike. X-ray-inspired imagery was eagerly appropriated by artists. The walls of courtrooms presented little barrier to entry, and x-rays were soon adopted as a key forensic tool, capable of finding hidden bullets still embedded in their unfortunate targets. X-ray machines eventually even found their way into neighborhood shoe stores, where one could ogle the bones of one's foot to pass the time. The health risks of prolonged x-ray exposure were not yet understood in those heady early days of x-ray fever.

The benefits of Röntgen's rays for health care, on the other hand, were appreciated quite early. Figure 3.3 shows an early x-ray image, once again

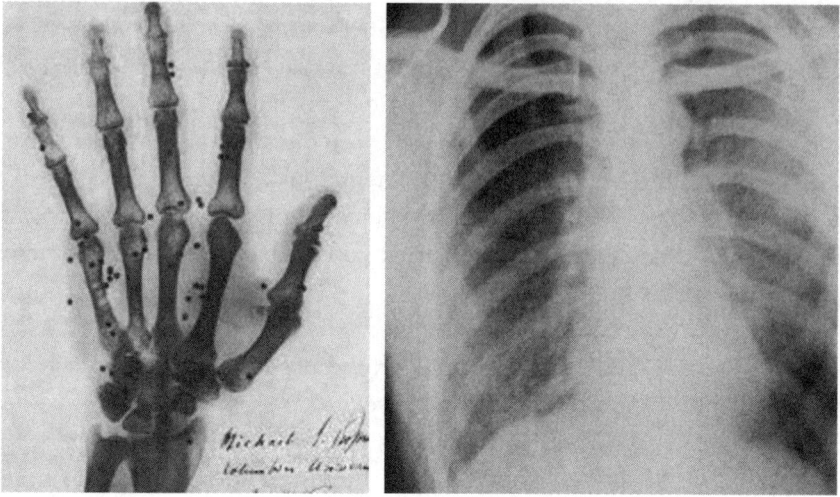

**FIGURE 3.3** Early medical x-ray images. (*Left*) An 1896 radiograph of a patient's hand, revealing pieces of buckshot. (*Right*) A chest x-ray film reproduced from a 1915 medical textbook.

*Sources*: (*Left*) Bern Dibner, *The New Rays of Professor Roentgen* (Burndy Library, 1963), 35; (*right*) Robert Knox, *Radiography, X-ray Therapeutics, and Radium Therapy* (Macmillan, 1915), 292.

from the banner year of 1896 and once again featuring a skeletal hand. This hand, however, is peppered with buckshot from a shotgun accident.[4] The dense pellets appear dark here because they block x-rays very effectively. This image was ordered by William Tillinghast Bull, a surgeon in New York who, like everyone else, had seen the famous Röntgen image. Bull figured that a similar image of his hapless patient would have value as an operating guide.

1896 was certainly the "year of the hand," but hands were just the beginning. Images of a wide range of body parts and medical disorders quickly followed. X-rays were used to ferret out all sorts of dense foreign bodies embedded in soft tissue, including bullets and other wartime shrapnel. The profession of dentistry was quick to realize the benefits of x-rays for the examination of teeth, which are even denser than the average bone. To this day, dental x-rays remain a routine part of life for many of us. X-rays would become the go-to tool for identifying and localizing bone fractures.[5] And then there is the chest x-ray (see figure 3.3), which can identify subtle changes in density associated with lung infections, for example. (Note that following what has become a medical convention, intensities in the chest x-ray image in figure 3.3 are inverted as compared with intensities in the adjacent hand image. Whereas bones appear dark in the x-ray image of the hand, dense structures like ribs appear bright and the air-filled lungs appear dark in the chest x-ray image.) The chest x-ray has become a kind of poster child for routine medical imaging. Ultimately, the medical use of x-rays launched the field of radiology, which in some circles is still referred to as Röntgenology.

The rapid pace of x-ray adoption was possible in part because x-rays were comparatively easy to produce. Amateur tinkerers around the world built their own x-ray machines, as did some professional tinkerers like Thomas Edison. In a story harking back to Galileo and Lippershey, Edison had a prototype of what he would come to call his x-ray "fluoroscope" up and running by the second week of January in 1896, four days after hearing about Röntgen's extraordinary findings.

In addition to capturing the eye, much as daguerreotypes had done before, x-rays also figured large in the public psyche. They came to embody aspects of the spirit of the age: Freud's invisible influences, and the casting off of Victorian modesty.[6] Freud, by the way, was a contemporary of Röntgen's. At the same time that Röntgen was experimenting with invisible rays, Freud was just beginning to chart the invisible forces that drive our psyches.

Why did x-rays provoke this remarkable upswelling of interest? I would argue that x-rays were so compelling by virtue of the simple fact that they made the invisible visible. What showed up on x-ray film was not just a replica of what we could already see. Instead, it was something beyond our eyes' natural capacity to capture. When trained upon the human body, x-rays revealed its secret inner mechanisms, and once we were privy to those mechanisms, we could never go back. The Copernican Revolution had changed how we saw our place in the cosmos. Likewise, the new form of transparency afforded by x-rays forced us to see ourselves in a new light.

Röntgen actually thought of his rays as "a new light." It turns out that this characterization was not merely a provocative analogy but was also rigorously correct. Like visible light, x-rays are electromagnetic waves. They just happen to have frequencies far beyond the visible spectrum. X-rays are light rays on steroids, blasting their way through things that easily block normal light. They are generated, we now know, when charged particles accelerate or decelerate rapidly—like when electrons in a cathode ray tube crash into dense metal and come to a sudden stop, or when hot gases swirl around stars or fall into black holes. Once generated, x-rays pack quite a punch. They are capable of knocking important pieces off delicate tissue, which is why we now wear lead vests during x-ray examinations; doing so minimizes incidental exposure.

The mystery of x-rays has faded over time, but their impact has continued to grow. In mastering x-rays, humans did with deliberation something that sensing creatures had evolved to do long ago: we used a new kind of probe to bring us new information about our world. As we shall see in the next chapter, many new imaging technologies to follow would have their origins in the discovery of a new probe.

In addition to constituting an important new probe, x-rays also introduced us to a new type of transformation. This particular transformation was akin to but strikingly different from the blocking or bending of visible light. A hand placed in front of visible light creates a shadow picture. X-rays—Röntgen's "new light"—can also produce shadow pictures, but with a significant twist. X-ray images are not simply shadows, but *projections*.

Projection is a concept familiar to most of us from maps and movies, in which three-dimensional objects are represented on a flat surface. An x-ray projection, however, is a little subtler. It involves squashing together the internal structures along the path of the x-ray beam so that what you see on the film or screen is a combination of all those structures, one on top of the other. The same thing happens when light travels through multiple partially transparent objects, like a table full of glassware. Wherever the light lands, you see multiple shadows superimposed: a drinking glass, a pitcher, a candlestick holder, all mingled together. The concept of x-ray projection is illustrated for a simple pair of objects in figure 3.4. Rays emerge in a beam from the x-ray source and pass through any objects in their way, eventually arriving at a sheet of film where their intensities are recorded. Two sample rays with different paths are shown schematically in the figure. Since x-rays are partially absorbed by every dense structure that they encounter along their path, the total absorption evident at any location on the film represents a sum over densities in each slice along the path traveled by the x-rays reaching that location. This means that x-ray images can provide excellent spatial discrimination in every direction *except* the direction of the x-ray beam. In that direction, we cannot tell here from there at all. For example, if the cylinder in the figure were

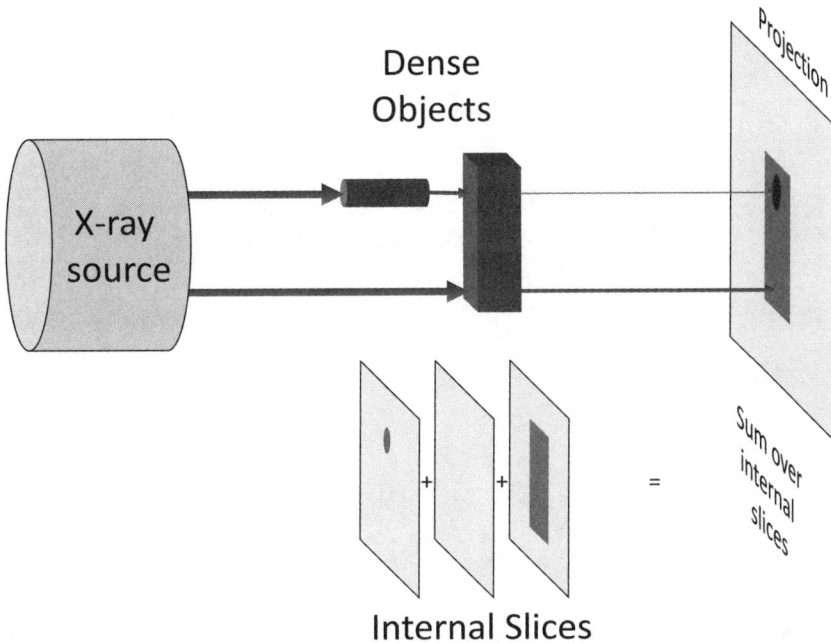

**FIGURE 3.4** A new transformation: x-ray images are projections of dense objects, meaning sums over internal slices along the path of the x-ray beam.

*Source*: Daniel K. Sodickson.

placed behind the rectangular block rather than in front of it, the resulting projection would be unchanged, and we would have no way of knowing that the objects had been switched.

Just as bends in our line of sight gave us magnification, x-ray projections gave us something extraordinarily powerful. For the first time in history, doctors armed with x-rays could identify objects both natural and foreign lodged within their patients' bodies. Because the shadows of those objects were superimposed on shadows of the rest of the body's structures, though, doctors couldn't yet nail down precisely where the objects were in three-dimensional space. That would take another type of transformation altogether.

# 4

## Slicing Without Cutting

How recording different views with various penetrating
probes allowed us to create cross-sectional images of the
insides of bodies.

**W**hen it comes to sorting out an elusive third dimension, we have a useful biological model to follow. I refer to the process of stereopsis, which contributes to depth perception, as discussed briefly in chapter 1. The position of objects along our line of sight is ambiguous in the same way that the position of objects is uncertain along the direction of an x-ray beam. Is an object far away and big or nearby and small? With only a single two-dimensional image to go by, and without knowing exactly how big the object is supposed to be, it can be difficult to tell. Fortunately, nature designed a convenient solution for us: it gave us two eyes. If we compare the views in these two separated eyes, we can extract useful information about depth. The apparent location of an object in the foreground of our vision—say, a pencil held in front of our face—shifts a lot between the view in our left eye and the view in our right. Light from the pencil comes into our eyes at significantly different angles, and it therefore strikes our two retinas in significantly different spots. A background object, on the other hand, such as a mountain in the distance, has very little apparent shift, since light from different spots on the mountain reaches our two eyes at nearly the same angle. Comparing views between our two eyes is not generally something we do consciously, but you can choose to make it conscious right now by closing one eye at a time. Do you see how the foreground shifts with respect to the background? This is a cue that our brain uses to give us a sense of depth. Our two eyes put the "stereo" in "stereopsis," helping us to judge distance along our line of sight just as our two ears allow us to distinguish the location of different sources of sound.

The analogous solution for penetrating probes like x-rays is called tomography, a word derived from Greek roots denoting "the writing of slices." And slices are precisely what tomography gives us—without the use of scalpels or other sharp implements! Practitioners of tomography routinely carve their way through delicate structures without a second thought. They inspect various cross sections of living beings, noting the precise location of familiar landmarks and looking for anomalies. They take human bodies apart, section by thin section, and put them back together again. The people subjected to this rigorous inspection then get up and walk away unharmed. How did we arrive at such a remarkable juncture in our capacity for visualization?

The trick to sorting out the internal slices buried within projections was actually figured out all the way back in 1917 by a mathematician with the singularly appropriate name of Radon.[1] Johann Radon appears to have had no particular interest in x-rays per se—he was simply intrigued by some of the puzzles that projections posed. His name lives on in the so-called Radon transform, a term still used to refer to various means of representing an object through its projections. When it came to imaging, though, the Radon transform was decades ahead of its time. It languished in the mathematical literature until the 1970s. Then, kaboom! The next imaging revolution was upon us.

The decade starting in 1970 saw its share of iconic figures and left a record of memorable images. Indira Gandhi. Margaret Thatcher. Mao Zedong. Muhammad Ali. Elvis Presley. Steven Spielberg. Steve Jobs. It was also the decade when cross-sectional anatomy revealed itself in sharp relief. While others were busy making music, or movies, or historic film clips, or personal computers, a small cadre of scientists was hard at work turning computed tomography (CT), magnetic resonance imaging (MRI), positron emission tomography (PET), and ultrasound into a practical reality. One pair of imaging innovators active at the time claimed a Nobel Prize before the decade was out; another pair had their turn in Stockholm several decades later; and others are still waiting. Together, these visionary vision-enhancers transformed the practice of imaging. In this chapter, I tell the story of the four key tomographic imaging modalities that, by the time 1980 came around, had laid the groundwork for a vast medical imaging enterprise that continues to this day.

◉

The key to tomographic imaging is to take multiple distinct but complementary projections of the same body—the equivalent of distinct views in different eyes. For example, you can shine x-ray beams on an object from a variety of angles. The result will be a set of shadow pictures that flatten the internal structures of the object along different directions. What is left, then, is the sort of logic puzzle that some people enjoy solving in their heads. Given a set of shadow pictures, what unique internal structures must have produced the various shadows? What shapes look this way when flattened left to right and that way when flattened front to back? This is the puzzle that tomography solves automatically.

To make things a little more concrete for you, figure 4.1 illustrates a simple approach to generating cross-sectional images from sets of projections. Let us say we wish to see the cross section of a simple object consisting of two dense cylinders suspended in air inside a round box with thin, low-density walls. First, as shown in the left-hand column of the figure, we send a beam of x-rays through the object from multiple distinct directions, all crossing the slice of interest, and we record the resulting projections. Since the x-rays pass through the object only in a single plane, the projections are not full two-dimensional images but just lines whose varying brightness reflects the profile of the dense cylinders seen from different angles. For example, the projection from below (top left of the figure) shows a bright bump for each cylinder; the projection from sideways-on shows only a single bright bump for the two overlapping cylinders; and the projections from intermediate angles show bumps with narrower spacing.[2]

The right-hand column of the figure shows how an image may be assembled from these projections. The procedure is known as back projection, and it is, almost literally, a means of "backing out" the contents of the object. As shown at the top right of the figure, the first projection is stretched out across the slice back the way the x-rays traveled. This back-projection process simply copies the brightness at each point along the line of the projection over the whole plane. This results in two bright vertical

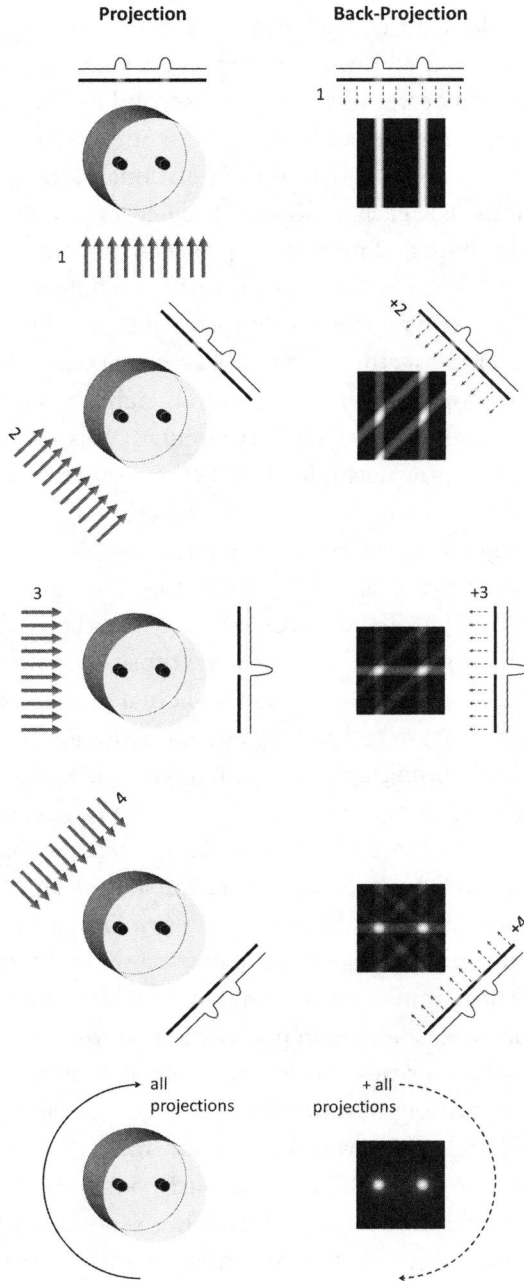

**FIGURE 4.1** The essence of tomography: creating cross-sectional images from projections.

*Source*: Daniel K. Sodickson.

stripes passing through the location of the dense cylinders. This may seem like an arbitrary thing to do, but in fact it makes some sense: since we don't yet know where the brightness originated, we assign it equally to all points along the path of the x-rays. For the second projection, we perform a similar back projection and add the result to our original result. Now we have two sets of stripes that cross at the positions of the cylinders and at two other locations. For each additional projection in turn, we repeat the process, adding in more and more back-projected stripes. Since the stripes overlap consistently only at the locations of the cylinders, the brightness accumulates strongly there, whereas it is spread thinly throughout the rest of the slice. If we add together enough back projections all the way from the bottom to the top of the object (bottom right of the figure), we get an image showing only the two distinct circular cross sections of our two cylinders. Our image now faithfully depicts the desired slice. If we're interested in a different slice, we just change the location of our projections and start the back-projection process over again.

That's it! Conceptually at least, that's how tomographic imaging is done.[3] We take enough distinct projections to characterize the contents of an object from its shadows, and we trace those shadows back to where they originated. It seems simple enough. In actual practice, though, it took some key developments in technology and computation to get us there.

The inventors of tomography didn't have the benefit of hindsight, of which I am making liberal use here. They weren't necessarily thinking about how to emulate the biological model of stereo vision. They didn't have the comfort of knowing that evolution had solved their problem for them. They did, however, fall back on a principle to which evolution had returned time and time again: for color vision, for stereopsis, and for various other sensory innovations. That key enabling principle, at least, is simple: when one view is not enough, take more views, and compare them.

Early imagers had already made use of such a principle to triangulate the location of foreign bodies using x-rays. The acclaimed radiologist Elizabeth Fleischmann was known for mentally piecing together internal structures from multiple judiciously positioned x-ray images.[4] Others devised clever mechanisms of moving x-ray tubes and x-ray film in opposite directions so that some internal slices stayed clear while other slices were blurred beyond recognition. Devices that took such an approach were even called tomographs, and they yielded messy but arguably recognizable cross-sectional images. As it would turn out, the key to cleaning up tomographic images lay in a marriage between mechanics and mathematics.

In the 1950s and 1960s, a particle physicist named Allan MacLeod Cormack, unaware of Radon's earlier work, had taken an interest in the localization of internal structures using x-rays.[5] His comprehensive theory of the mathematics of reconstructing objects from projections was eventually published in two parts in the *Journal of Applied Physics* in 1963 and 1964. To his seminal papers, as Cormack later stated in his Nobel lecture, "there was virtually no response."[6] Perhaps publication in a journal pitched to physicists rather than physicians didn't help. Neither, one might speculate, did the title he chose for his magnum opus (can you say "Representation of a Function by Its Line Integrals, with Some Radiological Applications" ten times fast?). Perhaps more to the point, though, just like Radon before him, Cormack was ahead of his time. He validated his theory in some simple test cases, but to realize the full potential of tomographic imaging with x-rays, he would have needed a reliable way to record a large number of reproducible projections along well-defined angles around a human body. This is what Godfrey Hounsfield brought to the table.

Hounsfield was a British electrical engineer working in the Central Research Laboratories of the transnational conglomerate Electric and Musical Industries (EMI), known for producing televisions, radar systems, and Beatles records. Having initially worked on guided weapons systems and radar, Hounsfield turned his attention to computers in the late 1950s, helping to build Great Britain's first commercial all-transistor computer. While on holiday in the countryside in 1967, he came up with an idea for figuring out what was hidden inside a black box by taking x-ray measurements at various angles around the box. On his return, he set about building a device to take the measurements. Starting with an improvised rig on a lathe bed, which took nine hours to complete a scan, he made rapid improvements until he could gather projections from a comprehensive set of angles in minutes or seconds rather than hours. For his imaging targets, he progressed from preserved postmortem human brains to fresh cow brains to his own healthy brain. He was certainly not afraid to get his hands dirty or to put his head down.

Another key component of Hounsfield's EMI prototypes was a computer to piece together images from a digitized set of measured projections. Here, Hounsfield's background in digital computing served him well, freeing him from the physical constraints of film and allowing him to retransform projections at will.[7] In our historical tour of imaging so far, we have already witnessed the power of transformations, including the bending of light and the shadow effect of x-rays. The computer opened up a whole new world of transformations. For the first time, an image was not merely what appeared on film or revealed itself to the eye. An image was what you made it. This new freedom to construct images from abstract sources, like recorded projections, set the new world of computed tomography apart from the old world of mechanical tomographs. The essential role of computing may also help to explain why cross-sectional imaging took so long to reach full bloom despite so many early seeds. It was waiting for computers that were up to the task.

On October 1, 1971, at the Atkinson Morley Hospital in Wimbledon, the world's first commercial CT scanner trained a series of x-ray beams on the head of its very first patient. Clinical workup had suggested that the woman had a brain lesion. Hours later, a computer had crunched the numbers, generating an image that clearly showed the dark circular cross section of a cerebral cyst. We may speculate about what the patient may have felt on learning that there was indeed something abnormal in her brain but that it was not a tumor. For his part, Godfrey Hounsfield must have been relieved. The scanner that was his brainchild had worked. Welcome to the seventies.[8]

CT ended up bracketing the decade quite nicely. In 1979, the Nobel Committee gave the nod to Hounsfield and Cormack, a mere eight years after the first clinical scan. On the surface, such a short turnaround time harks back to Röntgen. Admittedly, the advent of CT represented not the de novo discovery of a new physical phenomenon like x-rays but rather the creation of a powerful new tool using that phenomenon as a probe. Like the early x-ray machines, however, CT machines proved to be very useful tools indeed. Despite the dramatic complexity of a CT scanner as compared to an x-ray machine, CT was already being used in as many as one thousand hospitals around the world by the time of the 1979 Nobel ceremony.[9]

You know by now, of course, that the groundwork for CT had been laid long before. Since the Nobel Prize is not awarded posthumously, Radon missed out, as did various other contenders along the way whose contributions were not seen as sufficiently influential.[10] Setting aside the messy details of parentage, the prize was really recognizing the birth of tomography, after a long gestation.

Here's the kicker, though. Tomography doesn't work only with x-rays. You can use other types of projections as well. And in the early 1970s, some upstart scientists figured out how to make projections using magnets.

⟨❧⟩

Yes, I said magnets. I know you've borne with me so far as I've shared with you the story of humanity's progress in image-making, from probe to probe and from transformation to transformation. But do I really expect you to believe that magnets have the power to generate projections? I do.

Godfrey Hounsfield spent fully a quarter of his Nobel lecture on December 8, 1979, describing tomographic imaging not with x-rays, as he had devised it, but with magnetic fields. Cormack also felt obliged to mention the new phenomenon of magnetic resonance imaging. MRI was clearly the elephant in the room in the Stockholm Concert Hall that day.

On the face of it, magnetic resonance is nothing like x-rays. It involves the interaction of magnetic fields with tiny entities such as atomic nuclei (the small, hard bits of matter at the center of atoms). There are no rays to speak of. Magnetic resonance signals come from inside a body. They don't bounce off it or pass through it. So how in the world do you make an image with these signals?

In early September 1971, just weeks before the first successful clinical CT scan, Paul Lauterbur found a way. Lauterbur, a chemistry professor at New York's Stony Brook University, called his new method "image formation by induced local interactions." Here's how it works. When atomic nuclei like those in, say, hydrogen, are placed in a strong magnetic field and exposed to radio waves of a particular frequency, they emit an electromagnetic signal at the same frequency. Recall that your body is made up largely of water, which in turn is made up largely of hydrogen. So, if, for some unaccountable reason, you were to lie down inside a strong magnet and train a powerful radio antenna tuned to just the right frequency in your direction for just the right amount of time, you would become a living, breathing radio transmitter yourself. The nuclei inside the hydrogen inside the water inside you would emit a faint, transient radio signal.

So far, so good. But where's the image in that? If all the water in your body is emitting radio signals at the same frequency, what good are those signals in telling here from there? Here's the trick. Even in Lauterbur's day, it had been known for quite some time that the frequency of the magnetic resonance signal increases in direct proportion to the strength of the magnetic field. What Lauterbur realized was this: if you make the magnetic field strong on one side of the body and weak on the other, then the signal on the side with the strong magnetic field will have a higher frequency than the signal on the other side. So, just by analyzing the frequency of the signal you detect, you can determine where it comes from.

By placing a spatially varying magnetic field, known as a "field gradient," across a body, Lauterbur realized that he could create a one-to-one mapping between signal frequency and spatial position. In the presence of a magnetic field gradient, frequencies, and hence locations, are separated along the direction of the field gradient (say, left to right). Along other directions (say, top to bottom), however, there is no separation, and signals from different locations pile up together, just as they do along the path of an x-ray beam. In other words, in the presence of a field gradient, the magnetic resonance signal represents a projection of the body. Figure 4.2 illustrates this principle schematically.

Lauterbur had devised a means of creating projections with magnetic resonance. Next, he reasoned that he could rotate the direction of the field

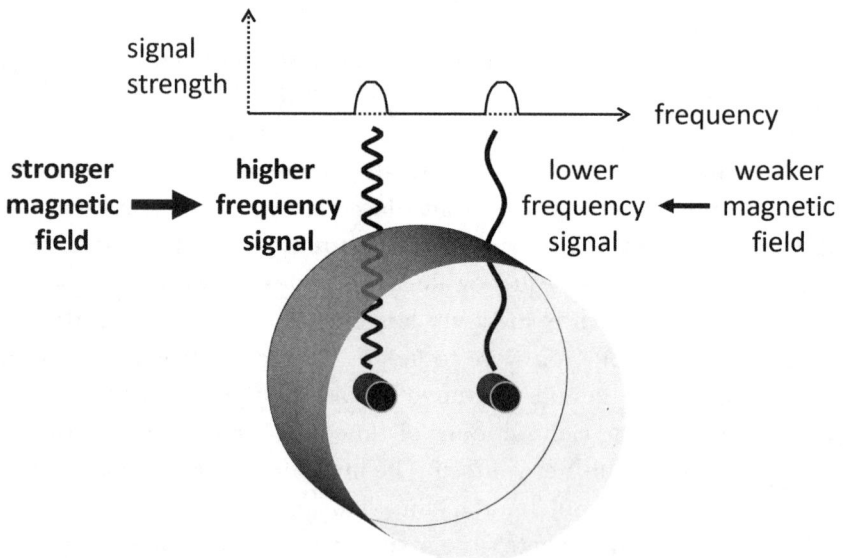

**FIGURE 4.2** MRI essentials: making a projection with a spatially varying magnetic field.
*Source*: Daniel K. Sodickson.

gradient and capture multiple distinct projections from different angles. He thought about the problem of reconstructing an image from those rotated projections and came up with . . . back projection![11] As it happens, he was unaware of the work of Radon, Cormack, Hounsfield, and other CT pioneers at the time, since the soon-to-be CT phenomenon was still under wraps at EMI. However, the first illustration in Lauterbur's seminal 1973 publication introducing MRI could easily be mistaken for an illustration of CT principles.[12] That publication in the prestigious journal *Nature*, by the way, was all of two pages long. A field founded in two pages. Not a bad day's work if you ask me.[13]

MRI was not yet ready for prime time, though. One particularly thorny practical problem involved how to choose your slice since, unlike x-rays, radio waves were exceedingly difficult to focus into a single plane. That's where Peter Mansfield came in. Mansfield was a physicist working at the University of Nottingham. A specialist in magnetic resonance, he had independently come up with a method of determining internal structure using magnetic field gradients. His paper on the subject was published in 1973, eight months after Lauterbur's landmark article appeared. Though it came at the imaging problem from a perspective very different from Lauterbur's, Mansfield's approach was still rooted in the separation of frequencies, and it shared the problem of distinguishing individual 2-D slices within the 3-D mix.[14] But in 1974, Mansfield and two of his students published a solution to that problem.[15] Their method of "selective irradiation" was yet another ingenious use of frequency separation. I won't burden you with the details here, but the net result was that there was now a way to image a single slice at a time, analogous to what could be accomplished with a planar beam of x-rays. At least in principle, all the key pieces for human imaging were in place.

Now that you know, in principle, how to make a magnetic resonance image of a human body, what do you think such an image will show? Unlike x-ray CT, it will not simply indicate the relative density of tissues. Only atomic nuclei attuned to the right frequencies generate any magnetic resonance signal at all. If you use frequencies associated with hydrogen nuclei, then you might expect your image to indicate the quantity of water (and other hydrogen-containing molecules) in any given region. Your liver and your brain have lots of water, and they will thus show up clearly in such an image. On the other hand, the air-filled spaces in your lungs and the calcium-filled surfaces of your bones are largely devoid of water and will thus be comparatively dark. So, at baseline, you would expect MRI to be good at depicting soft, squishy, fluid-filled tissues. This makes for a nice complement to x-rays, which excel at delineating hard, bony structures.

Another key difference between MRI and CT also results from the fact that MRI does not use x-rays. Nowadays, high-energy rays like x-rays are known as ionizing radiation because they have the capacity to knock small pieces off sensitive biological molecules like DNA. As a result, care must be taken not to use more x-ray exposure than is needed.[16] Not so with MRI. The radio waves used as a probe in MRI are still electromagnetic radiation—yet another "new light" that provided us new views of inner space—but their frequencies are far *lower* than those of visible light. This means that they have a whole lot less energy than x-rays. Radio waves at high enough power can generate heat in tissue, so the power is automatically controlled in MRI machines. Otherwise, the main safety considerations associated with MRI relate to the fact that it uses a very strong magnetic field that can exert a strong pull on certain metals and other materials. Operators of MRI scanners are trained to ensure that no one inadvertently brings magnetic objects into the vicinity of the magnet tube.

Speaking of the tube, you might wonder why an MRI machine is shaped like a tunnel even though there is no need to move x-ray sources or detectors around to get different projection angles. In MRI, the magnet is wrapped around you. The tube of a modern MRI machine is wound with miles of superconducting wire bathed in cooling fluid chilled to near absolute zero. It also contains radio antennas, generators of magnetic field variations, and other assorted electronics. I feel obliged to explain one last thing at this point, since it is the first thing most people who have had an MRI ask about: What causes the loud banging and buzzing sounds that assail you during a scan? Once again, this comes down to the field gradients that Lauterbur and Mansfield first employed to create projections. Each time the electric current passing through the coils of wire that generate magnetic field gradients is changed to switch to a new projection, the wires experience a force from the strong main magnet, resulting in a "clunk" as they react. When the gradients are changed quickly, the individual clunks string together into the characteristic, and ultimately harmless, droning sound that has become associated with MRI.

In fact, there is an art to how field gradients are changed and projections are played out, and a practiced ear may distinguish any number of subtle variants to the MRI drone. Starting in the mid-1970s, scientists figured out how to queue up different projections not just angle by angle but also in a wide array of other efficient ways that came to be known by fancy space-age names like "spin warp" imaging.[17] The creative ferment continues to this day. It hurts my nerdy heart, which I long ago gave to MRI, to skip over all these remarkable innovations and innovators—but the larger story beckons. Suffice it to say that by the time Hounsfield and Cormack were noting it in their lectures, MRI was off to the races.

Unlike for Hounsfield and Cormack, those races did not include a Nobel Prize for quite a long time. The prize for MRI was awarded only in 2003—three full decades after the key original publications appeared.[18] Why? Perhaps from the vantage point of the late 1970s, MRI looked like just another variant of CT since the two had not yet gone on to evolve quite separately. Perhaps it was a matter of speed: for various technical reasons, it took longer for MRI pathbreakers to produce human images than it had for CT, and the technology didn't get picked up by companies and introduced into clinical practice until well into the 1980s. A more likely explanation for the long Nobel delay, though, went by the name of Raymond Damadian.

Damadian was working at the State University of New York's Downstate Medical Center when he discovered that magnetic resonance signals from certain tumors lasted notably longer than those from healthy tissues. His signature 1971 paper documenting these findings in the journal *Science* laid out a powerful motivation for detecting tumors with magnetic resonance, and Damadian single-mindedly set about developing the technology to do so. There was only one problem—he didn't yet have a means of telling which parts of the magnetic resonance signal came from which parts of the body. The means he eventually dreamt up (which bore some resemblance to the old pre-CT tomogram approach of blurring out everything that one didn't want to see) have not proven to be nearly as powerful as what Lauterbur and Mansfield devised. Nevertheless, he did stake an early and well-documented claim to a kind of imaging with magnetic resonance.[19] When this claim was not broadly acknowledged in the emerging MRI community, Damadian vented his spleen publicly at Lauterbur in scenes reminiscent of the Lenard v. Röntgen affair. With the better part of a century of experience under its belt since Röntgen and Lenard, the Nobel Committee was none too fond of public conflicts over inventorship, and it kept its distance for many years. When the 2003 Nobel citation finally arrived, celebrating Lauterbur and Mansfield but including no mention of Damadian, Damadian's company, the Fonar Corporation, took out full-page ads in *The New York Times, The Washington Post,* and the *Los Angeles Times* protesting the omission. The topic still raises hackles today among those in the MRI community who lived through it. Regardless of anyone's feelings on the matter, this particular debate over credit does raise some interesting chicken-and-egg questions about which is more foundational as a concept: how an image is made, or what that image contains.

Our 1970s story of new images and new image-making is not yet done, though. While CT and MRI were generating buzz around the world, another tomographic imaging approach was coming into its own. If you thought taking pictures with mystery rays or magnets sounded a little like science fiction, try this one on for size: PET generates images using antimatter!

PET imaging is CT turned inside out.

As with CT, the source of projections in positron emission tomography is high-energy radiation, which blasts its way through tissue. But unlike with CT, the radiation tracked by PET scanners originates *inside* the body rather than being trained on the body from outside.

PET signal comes from injected "radiotracers." A radiotracer is a chemical tagged with a radioactive element. In PET, small quantities of radiotracer are introduced into the subject's body, generally via an easily accessible vein. The tagged chemical then makes its way through the body, following the course of blood vessels and collecting in certain tissues—for example, tissues that use the chemical in the course of their normal biological functioning. The radioactive element in any given molecule of tracer eventually decays, emitting a positron in a random direction.

This is where antimatter enters the picture. A positron is the antimatter counterpart of an electron—the tiny charged particle that orbits atomic nuclei and accounts for many of the chemical properties of matter. (It is also the electron that is the common carrier of electric currents in metal.) Once it has been generated by radioactive decay, the positron travels a short distance before inevitably encountering an electron, since electrons are just about everywhere in normal matter. Upon meeting its nemesis, the positron promptly "annihilates," immolating both itself and the hapless electron in a tiny explosion, like a supernova in ultra-miniature. This nano-nova in turn sends out two high-energy gamma rays in opposite directions. Gamma rays are electromagnetic radiation like x-rays but with even higher energy.

In a typical PET machine, as illustrated in figure 4.3, the subject is surrounded by a ring of gamma ray detectors. Somewhere around the ring, two detectors are hit by two oppositely directed gamma rays traveling at the speed of light away from an annihilation event, and nearly simultaneous signals appear in that pair of detectors. The PET machine looks for these so-called coincidences, or simultaneous, paired detector events. Once a pair of detectors has been identified in a coincidence detection, we know that the positron's last position before its demise lay somewhere along the line connecting those two detectors, since the gamma rays are known to have traveled in a straight line.

As dramatic as it may be for the subatomic particles involved, a single coincidence event like this isn't much help in making an image. However, events pile up over time as more and more radiotracer molecules in various positions throughout the body decay, emitting gamma rays along many

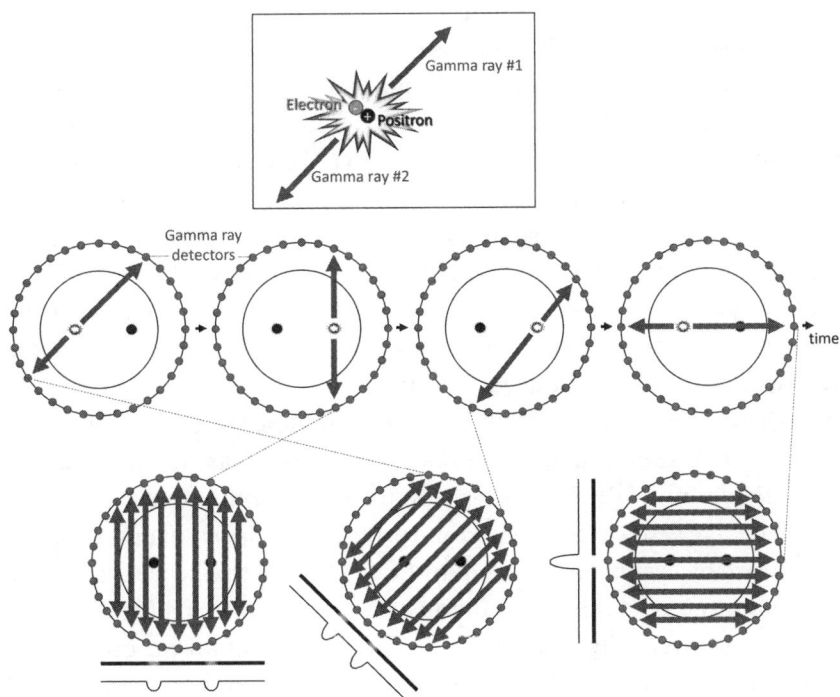

**FIGURE 4.3** PET essentials: making projections with antimatter and gamma rays.

*Source*: Daniel K. Sodickson.

distinct angles. You can then collect information from all the lines connecting detectors along any particular direction. This gives you a measurement of how much radiotracer lies along that direction. In other words, it gives you a projection of the radiotracer distribution. Gather projections from a full set of directions surrounding the body and—well, by now, you know what to do. You use your algorithm of choice to generate a cross-sectional image from the various projections.

PET developed in fits and starts as a labor of love by an eclectic group of investigators. Rudimentary localization of hidden radioactive sources was undertaken quite soon after radioactivity was discovered, but full-fledged imaging of radioactive tracers took its time getting going. The Hungarian physicist George de Hevesy leads a long list of explorers who began studying the behavior of radiotracers at the beginning of the twentieth century. This list includes Irène Joliot-Curie (the daughter of Marie and Pierre) and her husband, Frédéric, as well as the Americans Ernest Lawrence and Louis Sokoloff, among others.[20] There is also a rich history of localization algorithms for emission imaging. Cormack actually referenced emission methods using radioactivity in his seminal papers on tomographic image reconstruction, and the fields of CT and PET reconstruction have continued to be strongly linked over time. Tomographic machines using positron emission also developed gradually in the hands of inventors like Gordon Brownell, Benedict Cassen, Edward Hoffman, David Kuhl, Michael Phelps, Michel Ter-Pogossian, and Henry Wagner. A clinical PET scanner in something resembling its modern form did not appear until 1975, after Hounsfield's publications about his CT system prompted PET investigators to up their tomographic game. The now-standard radiotracer fluorodeoxyglucose (FDG) was introduced only in 1979 by Kuhl and his colleagues Joanna Fowler and Alfred Wolf. So, while a number of decades get credit for productive simmering, the 1970s reign supreme as the magic decade for the flowering of PET, together with its tomographic cousins.

Perhaps as a result of this shared and progressive history of innovation, no Nobel Prize has yet been awarded for PET.[21] Maybe that will have changed by the time you read this book. But maybe not. It remains difficult—in this case, uncontroversially difficult—to pin PET's invention on one, two, or the statutory maximum of three people who can share a Nobel Prize. PET also continues to have a narrower use in clinical practice than CT or MRI, though it is certainly the subject of continued vigorous interest for basic research in neuroscience, oncology, and many other fields.

PET differs from CT in that its gamma rays are *emitted* at random angles from within the body rather than being *transmitted* through the body at known angles from controlled x-ray sources. In practice, this means that PET scanners need to wait until enough random decays have been registered in the gamma ray detectors to get complete projections along a sufficient number of directions. It also means that the radiation dose considerations for PET are a little different from those for CT. Gamma rays are ionizing radiation, just like x-rays. In PET, though, the radiation is tied to the tracers, not to the scanners per se. Radiation dose accumulates not just when a scanner is gathering projections (as with CT) but for as long as the injected radiotracer remains radioactive inside the body. Fortunately, small quantities of radiotracer with a relatively short half-life are generally sufficient to generate good images, since the PET detection mechanism is highly sensitive. PET also involves the injection of a chemical substance that carries some risk of allergic reaction or other side effects. Bear in mind, though, that the most common radiotracer used in PET scans nowadays is a simple variant of glucose—that is, common sugar—which most of us consume in large quantities without explicit medical justification.

Since the probe that generates projections in PET differs from that used in CT or MRI, PET images also yield distinctive information about the body. Rather than showing the full distribution of internal structures, PET images indicate only where radiotracers concentrate. If you use a glucose-based tracer for a PET scan, for example, you will see hots spots on the image wherever glucose is most needed for biological processes. In other words, you will see a map not of anatomy per se but of sugar *metabolism* in tissue. As a result, PET is considered a form of *functional* imaging, in the sense that it depicts not just tissue structures but tissue functions such as metabolic activity.[22] Modern scanners actually combine PET with CT (or, in some cases, MRI) to provide key anatomical context for regions of high tracer uptake.

One type of tissue with particularly vigorous metabolic activity is cancer. Not surprisingly, then, PET is now used routinely to detect cancer. Whole-body PET screening studies take advantage of the high sensitivity of PET to small concentrations of radiotracers in order to identify regions of anomalous metabolic activity that might represent budding tumors. PET is also used to track the progress of cancer treatments, since successful chemotherapy, for example, can shut down metabolism before substantial changes in anatomical structure become visible. Another key area of application for PET lies in functional imaging of the brain, which is known for its lively metabolic activity. The brain is also awash in important neurotransmitter molecules that can be targeted by designer radiotracers that bind avidly to particular chemical compounds. PET is therefore a prime example of what is now called *molecular* imaging.[23]

The affinity of PET for particular molecules of interest has a powerful analogue in the deep evolutionary history of imaging. I refer here to our senses of smell and taste. Our noses and our tongues are exquisitely sensitive to small quantities of particular chemical compounds. In this sense, PET is like an artificial nose—albeit one that uses principles of tomography to map out the locations of whatever it sniffs out. This analogy raises the question of what other kinds of natural senses we might try to emulate. What about hearing, for example?

Bats and dolphins figured out how to make images with sound long before humans ever did. These creatures take advantage of a process called echolocation. Using the various vocal mechanisms they possess, they send out a ping (or a chirp or a screech), and they listen for an echo. The time delay between the ping and the echo reveals the distance between them and whatever structure caused the ping to bounce back. The farther away a cave wall or potential prey animal is, the longer it takes for sound to travel out and back. In other words, the time it takes for an echo to return to its sender has a direct correspondence with distance. Thus, it is possible to create a local map of the world using sound as a probe.

Humans eventually realized this fact and set out to emulate what came naturally to bats and dolphins.[24] After an undetected iceberg sank the Titanic in 1912, and after stealthy submarines started sinking surface vessels in World War I, the problem of protecting ships with sound-based echolocation was taken up. These efforts culminated in the development of sonar (which stands for "sound navigation and ranging") during World War II.[25]

Echolocation allows one to determine distances to sound-reflecting surfaces like the trunk of a tree or the hull of a submarine. How, though, can one reach inside? How can one create cross-sectional slices of the sort to which you are now accustomed? How can one do tomography with sound?

It turns out that sound doesn't just reflect off the surface of the human body. Some of it is also transmitted though body tissues. Sound, after all, is a traveling mechanical vibration, and it can travel in any suitable elastic medium. So, you can send sound waves *through* the body and listen for any echoes that bounce off structures along their path. To do this effectively, you need sound not only to travel well through the body but also to reflect well from internal structures. Audible sound is effectively transmitted through body tissues (if you put an ear or a stethoscope to someone's chest, you can hear their heartbeat, after all), but sound waves in the frequency range that humans can hear have comparatively long wavelengths (the distance from one peak to the next), and they pass right through small body structures without reflecting strongly. It makes sense, then, to use ultrasound, which is just sound at frequencies higher than we can hear. With wavelengths of a millimeter or less, ultrasound can be transmitted through reasonable depths of tissue while still reflecting effectively from millimeter-scale body structures. In other words, ultrasound is an effective penetrating probe.

Both bats and dolphins can produce and hear sounds that extend well into the ultrasonic range, but we humans require technological help to do so. A key enabling technology for generating ultrasound waves artificially is based on the so-called piezoelectric effect, discovered by none other than Pierre Curie and his brother, Jacques, way back in 1877, before Röntgen discovered x-rays and before Pierre discovered Marie. By applying an alternating electric current to a piezoelectric crystal, you can generate pressure waves—that is, sound—with the same frequency as the applied current. Likewise, when a piezoelectric material is compressed by sound waves, it generates an electric current at the same frequency. With piezoelectric crystals, you can generate carefully controlled sound without vocal cords or other manual instruments, and you can also detect sound, or ultrasound, without ears.

Another pair of brothers—Karl and Friedrich Dussik—attempted to use sound to image the human body before World War II was over, but they did so using only transmission, treating sound like a poor cousin of x-rays. After the war, multiple groups started working to apply sonar spin-off technologies, and echolocation principles, to the generation of ultrasound images of internal body structures. The subsequent history of ultrasound development is arguably even more diverse and diffuse that that of PET, complete with all the strokes of genius and interpersonal conflicts that characterize innovative endeavors.[26] Then the rising tide of computer-aided tomography in the 1970s lifted ultrasound to new visibility as well.

Figure 4.4 illustrates how you can make a projection with reflected sound waves. You send an ultrasound pulse through a body of interest and carefully record the times at which reflected pulses arrive back at the source. (The gray "T" structure in the figure represents an ultrasound transducer that emits ultrasound pulses and also detects echoes of those pulses. Triplets of straight parallel lines represent schematic wavefronts of sound waves traveling in the forward direction, whereas triplets of concentric curved lines represent reflected sound waves traveling back toward the transducer.) Each distinct internal structure along the path of the pulse will show up as an echo with a distinct return time. All structures hit by the pulse at the same time, however, will be lumped together. So a plot of signal strength versus time delay will, in fact, represent a projection. There is a direct analogy here with MRI: just replace "time delay" with "frequency," and the signal plot at the top of figure 4.4 is more or less the same as the plot in figure 4.2.

One key difference between ultrasound and MRI is that ultrasound waves can be focused much more tightly than radio waves, so there was no issue with selecting a particular ultrasound slice. Quite the contrary: with the right technology, ultrasound can actually be focused all the way down into a narrow pencil beam. This beam can then be swept from side to side, and the image can be assembled line by line rather than just projection by projection. Another difference is that ultrasound transducers can be made quite small, and many transducers can be bundled together into a compact array. Figure 4.4 illustrates an idealized projection captured by a single extended transducer, but most modern ultrasound machines use many-element transducer arrays that can be coordinated to transmit sound waves together in desired patterns (whether broad wavefronts or narrow beams). Those same arrays can then listen out for individual echoes arriving at each element. The resulting plethora of individual projections can be combined to reconstruct images without ever needing to sweep a physical beam. The mathematical transformation which accomplishes this reconstruction is called beam-forming. All these features give ultrasound a significant advantage in speed and convenience over its tomographic cousins. You don't even need to surround the body as long as the echoes produced by incoming sound waves extend across your slice of interest. Like other tomographic imaging modalities, ultrasound began its life in large tubes. Some test subjects were even immersed in repurposed B-29 super bomber gun turrets filled with water to enhance sound transmission and match tissue densities. However, ultrasound was the first of the major tomographic modalities to escape the confines of the tube. In part II, we will explore which other imaging modalities may one day follow suit.

In the meantime, let's spend a moment on the question of what ultrasound images show us. From what I have described so far, you can deduce that the brightness of an ultrasound image at any given position depends

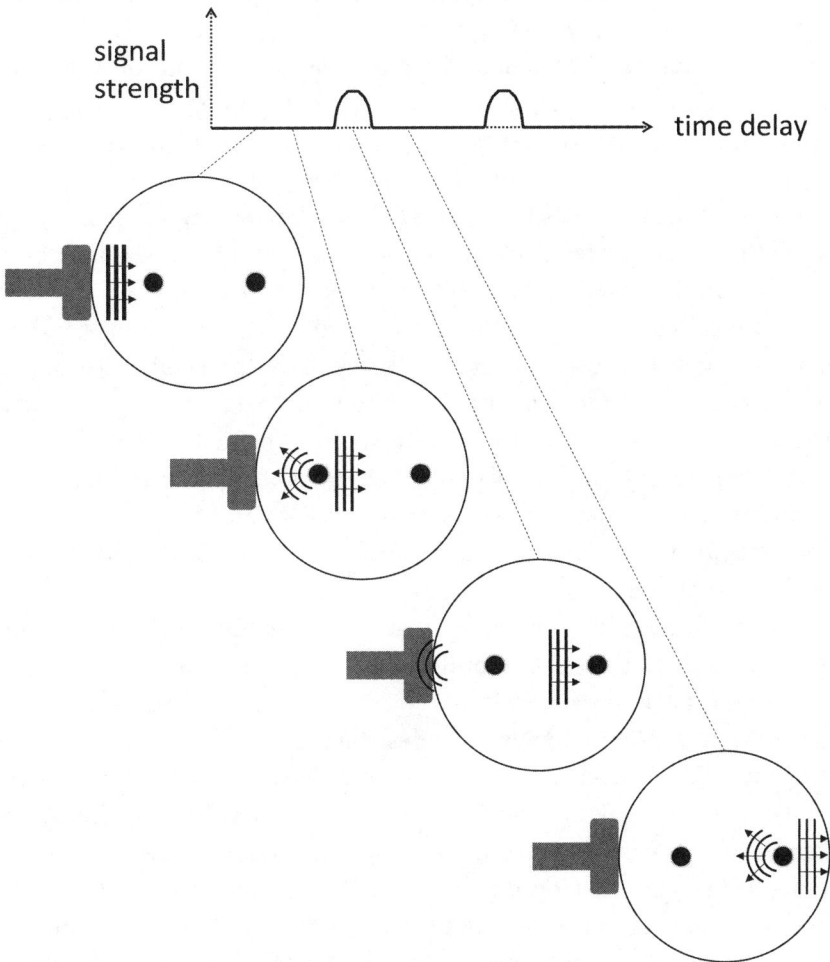

**FIGURE 4.4** Ultrasound essentials: making a projection with reflected sound waves.

*Source*: Daniel K. Sodickson.

on how much ultrasound signal bounces back from that position at any given time. As with CT, this is affected by the local tissue density. Unlike CT, though, ultrasound is mostly sensitive to *changes* in tissue density. Wherever the density changes, there is an interface from which sound reflects.[27] Changes in the elasticity of tissue also yield reflections. The many small variations in typical body tissues actually result in a profusion of little echoes that fill in the spaces between edges and give ultrasound images their notoriously fuzzy texture.

Many of what would become the most compelling uses of ultrasound are connected to its speed. While other tomographic imaging modalities must push their speed limits to capture motion, ultrasound routinely shows images being made in real time. It is widely used in cardiology because it can image the beating heart. It is also famous for imaging pregnant mothers and their precious (and sometimes restless) cargo. Since it is fruitless to ask a fetus in the womb to stay still while you take pictures, a fast camera is called for. Because the uterus is a fluid-filled structure that conducts sound well and because fetal bones, particularly the skull, have not yet hardened enough to block sound out entirely, prenatal imaging represents an especially effective use case for ultrasound. The field of obstetrics has certainly recognized the appeal of ultrasound and has embraced it as a part of routine prenatal care. In fact, ultrasound is one of the few cross-sectional imaging technologies to have significant reach outside the specialty of radiology (though radiologists also use it regularly).

Another feature of ultrasound that makes it appealing for imaging pregnant mothers is its safety profile. Like the radio waves used in MRI, ultrasound is a comparatively low-energy nonionizing probe. At moderate intensities, it is essentially just harmless high-frequency buzzing. On the other hand, highly focused ultrasound does have the capacity to disrupt tissue selectively. This has made ultrasound a promising tool for image-guided therapy, in which malfunctioning brain regions, liver tumors, or uterine fibroids are destroyed by focused ultrasound beams. In fact, other penetrating probes like x-rays and gamma rays, which are more difficult to focus than ultrasound waves, can also be used for targeted therapy using principles similar to those that allow cross-sectional imaging. In certain forms of radiation therapy, carefully-metered radiation is applied from distinct angles so that lethal doses accumulate only in specified body regions. While low-intensity applications of penetrating probes allow us to see unhealthy structures like tumors inside the body, carefully shaped, high-intensity applications of those same probes can actually allow us to do something about what we find—once again, without making a single cut.

Here ends our survey of the tomographic imaging modalities that took the world by storm in the 1970s and beyond. You now have a working knowledge of how to make cross-sectional images from projections. You have also encountered various ways to make projections using x-rays, magnetic fields, gamma rays, and ultrasound. I have introduced you to an extraordinary community of inventors who were responsible for the flowering of medical tomography. You have heard about some of their aha moments and also about some of their bitter disagreements. You have encountered world-spanning collaborations as well as solitary geniuses, distinguished Nobel laureates as well as unruly Nobel Prize controversies that played out on the pages of daily newspapers.

You didn't think I would tell you all this and not show you any actual images, did you? Figure 4.5 provides a montage of selected innovators, early machines, and seminal images for CT, MRI, PET, and ultrasound. More examples may be found in the color insert.

It does not escape me that most of the inventors' faces staring out at you from this photo montage are those of white men. This is consistent with the recorded history of scientific advances during the era in question. Fortunately, persistent ideological debates aside, there are signs that the world is beginning to wake up to long-standing inequities in both access to and recognition for careers in science. Going forward, I will do my best to call out a wider variety of faces and voices in the ongoing story of imaging. In chapter 7, for example, you will learn how the practice of medical imaging expanded to encompass large swaths of the human population, creating new careers in many walks of life. In the meantime, I apologize for the omission of any deserving but hidden figures in our whirlwind history of imaging so far.

As with any birth, the birth of tomography was just the beginning. Entire generations of improvements were in store for CT, MRI, PET, and ultrasound imaging. The basic principles of tomography, meanwhile, would find application with a wide range of other penetrating probes. Electrical impedance tomography (EIT) probes the interior of bodies using electric currents. Magnetoencephalography (MEG) converts tiny magnetic fields produced by the brain's intrinsic electrical activity into images. PET is only the most widely known representative of a broader field of nuclear imaging modalities, of which single-photon emission computed tomography (SPECT), which localizes different gamma ray-emitting radiotracers without coincidence detection, is another notable example. Principles of frequency encoding similar to those employed in MRI are at the root of x-ray crystallography (which is used to suss out the configuration of tiny molecules) and radioastronomy (which has recently given us our first up-close

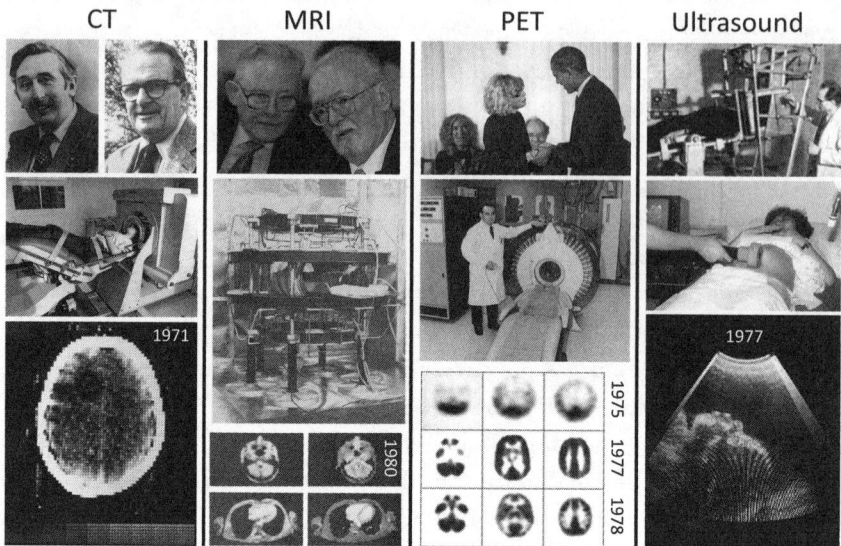

| CT | MRI | PET | Ultrasound |
|---|---|---|---|

**FIGURE 4.5** The birth of tomography: a storyboard. Selected snapshots of inventors, prototypes, and early images are shown for CT, MRI, PET, and ultrasound.

**CT:** Godfrey Hounsfield (*top left*); Allan Cormack (*top right*); Hounsfield's prototype EMI CT scanner, 1971 (*middle*); first clinical CT scan showing a brain cyst, October 1, 1971 (*bottom*)

**MRI:** Paul Lauterbur and Peter Mansfield (*top*); late-1970s prototype MRI machine in Aberdeen, Scotland (*middle*); early brain and body images from the Aberdeen scanner, 1980 (*bottom*).

**PET:** Joanna Fowler, a co-inventor of the PET radiotracer FDG, receiving a U.S. National Medal of Science (*top*); PET innovator Michel Ter-Pogossian beside a 1977 prototype scanner (*middle*); selected brain PET images from the 1970s (*bottom*).

**Ultrasound:** Karl Dussik with his early ultrasound apparatus, 1946 (*top*); late-1970s ultrasound device with a rotating handheld transducer (*middle*); fetal ultrasound image created using this device (*bottom*).

*Sources*: **CT:** (*top left, top right*) Wikimedia Commons; (*middle*) © Science Museum Group; (*bottom*) German Röntgen Museum, Remscheid. **MRI:** (*top*) Reuters/Bridgeman Images; (*middle*) reproduced with permission from J. Mallard et al., "In Vivo N.M.R. Imaging in Medicine: The Aberdeen Approach, Both Physical and Biological," *Philosophical Transactions of the Royal Society of London B* 289, no. 1037 (1980): 519–30, figure 5; (*bottom*) reproduced with permission from W. A. Edelstein et al., "Spin Warp NMR Imaging and Applications to Human Whole-Body Imaging," *Physics in Medicine and Biology* 25, no. 4 (1980): 751–56, figure 2. **PET:** (*top*) Courtesy of Brookhaven National Laboratory; (*middle*) courtesy of the Bernard Becker Medical Library Archives, Washington University School of Medicine; (*bottom*) CTI PET Systems. **Ultrasound:** (*top, middle, bottom*) Courtesy of Joseph S. K. Woo. (https://www.ob-ultrasound.net/index.html)

look into the distant maw of a giant black hole). Geologists and seismologists use tomographic imaging with vibrations to probe the invisible anatomy of Mother Earth herself. Tomography has even come full circle right back to visible light—the age-old model for all imaging probes. In diffuse optical tomography, cross-sectional images are formed using the residual light that manages to make its way through turbid tissue. Frequency encoding has been used for high-resolution optical microscopy. Optical imagers have even found a way to copy ultrasound. Optical coherence tomography (OCT) uses the minuscule but measurable transit times of reflected light to map out spatial positions in high-definition images. OCT has found medical use in ophthalmology to image—of all things—the human eye!

Speaking of the eye, our catalog of cross-sectional imaging probes brings us back to another recurring theme of this book. We see once again that imaging has thrived by emulating naturally evolved senses (whether our own or those of other species populating our planet's rich ecosystems). Each suitable probe that humankind has discovered has eventually been transformed into the equivalent of a new sensory modality.

Transformation is of course another recurring theme. Technology has allowed us to harness all manner of new probes, but it is how we transform the signals from those probes that ultimately allows us to see things in a new light. In the case of tomography, the transformation in question is a reconstruction of internal body structures from projections. Mathematically speaking, reconstruction from projections is one example of a general category of transformation called an inverse transform. Inverse transforms undo the effects of an original transformation like a projection.[28] The inverse transforms at the root of tomography are more abstract than the bending of light or the projections offered by x-rays. Whereas both magnified optical images and simple projection images make inherent sense to the naked eye, the ordered sets of projections that constitute the raw materials of tomography don't look anything like images we are used to seeing. It takes significant processing, generally accomplished with computers, to liberate the spatial information hidden within these raw projection signals. Historically, while this progressive abstraction has moved the mechanisms of imaging ever further from the public eye, it has also resulted in progressively increasing power. The future of imaging, as you will see, will likely hinge on still more powerful transformations, emulating not just our senses but our cognition itself.

One additional theme of this book involves societal impact. It was already clear to many observers by the end of the 1970s that tomography would have a transformative effect on medicine. A survey of doctors in 2001 validates this intuition.[29] More than two hundred leading general internists were asked to rank the relative importance of thirty medical innovations introduced during the previous thirty years. The winners, by a large margin, were MRI and CT. More than 75 percent of responding doctors indicated that, of the innovations they were presented with, MRI and CT were most essential to their patients' welfare. The competition for most indispensable innovation was nothing to sneeze at: coronary bypass surgery and angioplasty were on the list, as were statins and hip and knee replacement, among many other staples of modern medicine. In the end, though, eyes won out over drugs, blood tests, and surgical implements. Of all the medical miracles they were asked to choose from, the doctors were least willing to part with tomography.[30]

This vote of confidence reflected the dramatic impact that cross-sectional imaging had made on the day-to-day practice of medicine since the 1970s. Before routine tomography became available, patients with mysterious symptoms were often left with no alternative but exploratory surgery—a jaunty-sounding name for a decidedly unpleasant expedient that amounted to "let's open you up and see what we can see." As soon as robust cross-sectional imaging became feasible, however, researchers set out to catalog the visible signatures of internal pathologies so that cutting could be reserved for when it was truly needed. Anatomical exploration increasingly became the domain of comfortably unintrusive radiologists. Huge enterprises sprang up to feed radiologists and other clinicians with images. Many of the major multinational electronics companies—General Electric, Siemens, Philips, Hitachi, Toshiba—spawned new divisions devoted to cross-sectional medical imaging. Imaging became big business. Whole new professions came into being within the fast-paced technological and economic ecosystem of medical imaging as it expanded to include skyrocketing numbers of people. You will hear more about some of these people and their professions in chapter 7.

Not only did professional tomography affect the well-being of everyday people, but it also managed to get into the heads of the general public. Even if the workings of imaging machines were becoming increasingly mysterious to the lay public, all this virtual slicing and dicing still had a profound impact on our collective body image. Fetal ultrasound changed our conception of pregnancy, opening up the stages of prenatal development for all to see. While ultrasound was not great at penetrating fully developed skulls,

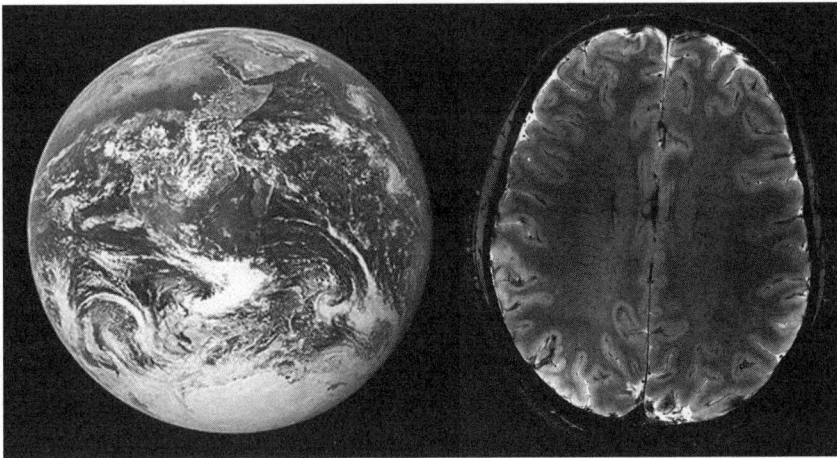

**FIGURE 4.6** Transformative images of outer space and inner space. (*Left*) *The Blue Marble*. (*Right*) The brain revealed by MRI.

*Sources*: (*Left*) Wikimedia Commons; (*right*) courtesy of Martijn Cloos and Graham Wiggins.

the skull was no obstacle to MRI, CT, or PET, and the resulting images had a similarly profound effect on our relationship with our brains. The x-ray projection of a typical head was not very interesting, but the new cross sections gave us front-row seats to the brain's inner workings—both its structures and, increasingly, its functions. Consciousness now had a visible home.

When it comes to psychological and societal impact, it is tempting to compare cross-sectional brain images with another landmark 1970s-era image which has come to be known as *"The Blue Marble"*—the photo of Earth taken by the Apollo 17 crew on December 7, 1972, as they made their pioneering way to the moon (figure 4.6). Just as *The Blue Marble* gave us new perspective on the little round rock that was our post-Copernican home, brain images showed us, for the first time in history, the little round space in which our innermost thoughts lived. Before these images were made, we had to imagine what the surface of our planet or the inside of our head might look like. Now, when you think of Earth, chances are you think of a blue marble in space. When you think of a living brain, you may well call to mind some of the brain images you've seen. What these images showed us starting in the 1970s could not be unseen. Whether we were looking through imaging tubes or lying inside them, whether we were surveying outer space or inner space, these images forced us to see ourselves in a new light.

# 5

# What's in an Image?

Wherein we consider the nature of images in general and explore the basic structure and information content of modern images, whether they emerge from cameras, microscopes, telescopes, or medical imaging devices.

By the end of the 1970s, we humans had at our disposal a dizzying array of distinct types of images. For the first time, moreover, some of these images had begun to take a digital form—as electronic bits of pure information. Before we continue with the next stage of our imaging story, let us take a moment to take stock of the diverse information that can be packed into modern images.

Generally speaking, such images share a common underlying structure, whether they are the result of photography, microscopy, astronomy, or tomography. They are characterized by common metrics of sharpness and image quality, and they are composed of small, irreducible bits like atoms. Within the confines of these common structural features, however, modern images can reflect all manner of distinct facets of our world. Photographs allow us to see in infrared or ultraviolet, translated into colors our eyes can comprehend. Telescopes span the spectrum from radio waves through x-rays to gamma rays, providing complementary pictures of our cosmic surroundings and ultimate origins. Medical images have a similarly wide range of information content, depending on which probes are used to interact with the body and generate projections. Even within a single imaging modality, there can be a remarkable diversity of information. In MRI, for example, the same anatomical feature can be made to appear bright, dark, or anything in between, depending on what you need to know about it.

This chapter begins by considering what constitutes an image from the vantage point of the late twentieth century. It concludes with the story of how a generation of creative thinkers tailored imaging methods to visualize everything from the birth and death of stars to the firing of neurons as we think.

What comes to mind for you when you think of an image? A favorite work of art, perhaps? Or a treasured photograph? In common parlance, *image* may refer to any appearance or perception, true or false. The word *image* has its roots in the Latin "imago," meaning "imitation." This original meaning suggests that some truth is being copied to create an image. But what does an image imitate? Let's try to be a little more precise.

I have asserted in previous chapters that the concept of an image encompasses any visual representation of spatially organized information. Such a definition is useful as a catch-all, but it is relatively nonspecific. Cave paintings, road maps, children's drawings, newspaper pages—they all qualify as visual representations of spatially organized information. A more concrete understanding of human-made images will require us to consider at least three things: (1) the source material from which a given visual representation is derived; (2) the means by which a representation is created from the source material; and (3) how spatial information is organized in the representation.

First, let us consider the source. In everyday lingo, we tend to describe images as being images *of* something: "This is an image of my cat." "That is an image of your spleen." This common intuition aligns with the original meaning of an image as a copy and suggests that the source of an image matters.

From what sources, then, can an image be derived? There are conceptual sources, such as an imagined scene or an abstract idea. There are also objective sources, such as a physical object or scene in the real world. The fact that all "real" scenes and objects come to us only as cognitive representations delivered through our senses—a fact not lost on a long line of philosophers—tends to blur the distinction between conceptual and actual. Nevertheless, it is fair to ask how faithfully an image represents its source material. Is the image a direct copy of some aspect of the source, or has artistic license been taken? A key feature of medical images, to take just one example, is that they are designed to help doctors understand what is really wrong with a particular patient. Artistic license is rightfully frowned upon!

That said, it is important to remember that images are *representations* of their sources, not the sources themselves. They all involve a process of rendering. In other words, all images are artificial. Like abstract paintings or distorted maps, even images based on precise measurement involve a kind of artifice. As such, they are subject to errors, which are often referred to rather evocatively in the field of imaging science as artifacts. What we see is definitely *not* always what we get. Remember this.

The question of artifice brings us to the means of creating visual representations. Different methods result in different representations and are associated with different types of artifacts. You have seen this play out in the history of imaging methods that we have explored together so far. As humankind discovered different probes of the world around us, we used them in various ways to understand the world and our place in it. Light was originally transmitted straight to our eyes until we learned to bend it. Hand-copying was replaced by automatic rendering using machines and algorithms. Direct correspondence between scene and image in cameras and optical telescopes gave way to increasingly complex transformations like projections and tomography.

At each of these stages, the organization of image information underwent subtle shifts. When our eyes were all we had, image organization was dictated by the optics of those eyes, the layout of cells in our retinas, and the interconnection of visual centers in our brains. In handmade images, the geometry of the canvas, the size of the brush, and the skill and intent of the artist dictated what was delivered to our eyes. With the advent of photography, brush and canvas gave way to camera optics and film. The density and distribution of silver grains in film, together with key lens characteristics, determined what the sharpest possible photographic image could be. Likewise for x-rays, captured for us by film grains or phosphor dots. The distribution of exposed grains or glowing dots determined how much you could zoom in before an x-ray image got unhelpfully blocky.

And what about tomographic images? How are they organized?

Well, in tomography, the organization of an image is actually up to us. This is a natural outcome of the progressive abstraction and digitization of imaging. The raw measurements underlying a tomographic image no longer correspond directly with the visual appearance of an object. A modern tomographic image is not simply captured on film or projected on-screen. It is recorded digitally, transformed by computer into a suitable visual representation, and only then displayed to our eyes, typically on a computer screen. And who tells the computer what to do? We do. We choose the parameters of the transformation that turns measurements into visual representations based on what our probe of interest can show us and on what we want to see. In an ironic twist that has always appealed to me, the ever-more-precise science of imaging has become a new kind of digital art. Even photography, which once set the standard for what-you-see-is-what-you-get imaging, has followed suit and gone digital. Nowadays, even straightforward-looking photos are as often as not the outcome of significant computation.

I do not mean to suggest here that the art of modern imaging is a free-wheeling creative exercise in which imagers can simply make up what they want to see. Quite the contrary. Even though all images are in a way fictions, they are not arbitrary fictions. An entire field of mathematics known as image reconstruction is devoted to the rigorous transformation of diverse measurements into organized images. I count myself as an enthusiastic member of the far-reaching scientific community that has sprung up around image reconstruction. Work by this community has laid out rules of the road for both gathering and transforming imaging data. If you want your images to have certain properties, your measurements must be collected in certain ways, and your transformation algorithms must include certain features. You can ignore these rules if you like, but you do so at your peril.

All this talk of image reconstruction is getting a little abstract, so let's make it nice and concrete. Figure 5.1 shows examples of raw measurement data and corresponding reconstructed CT, MRI, PET, and ultrasound images. Raw CT and PET data look like complex gray braids. Raw MRI data more closely resembles a starburst. Raw ultrasound data, perhaps unsurprisingly, looks like a tracing in an audio file. These patterns are seldom seen by patients or even physicians, but they can be a little mesmerizing if you stare at them long enough. Like a kind of visual DNA, these patterns contain the coded information that ultimately generates entire bodies rendered in cross section. For several years running, I welcomed a class of visual artists to the imaging center I directed at New York University. The purpose of this annual field trip was to show artists in training how medical images are made, and they were inevitably fascinated by the suggestive patterns in raw data.

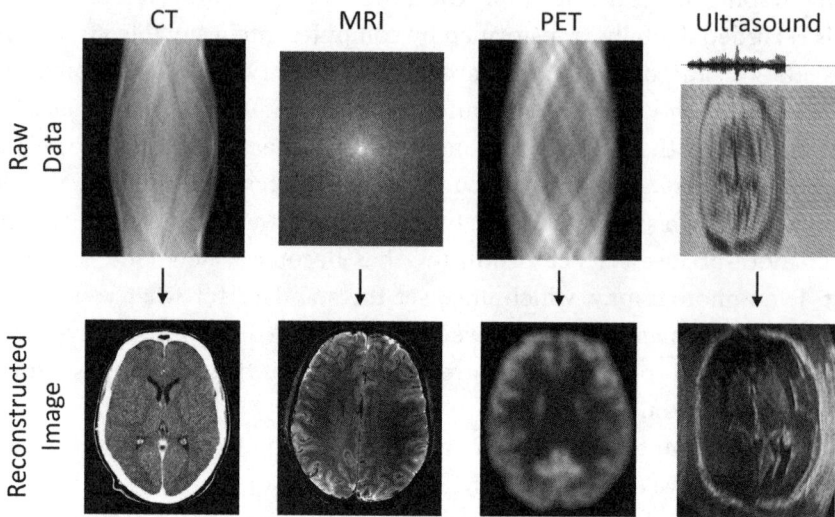

**FIGURE 5.1** The art and artifice of modern imaging: diverse sets of raw data are transformed into CT, MRI, PET, and ultrasound images.

*Sources*: **CT**: (*raw data*) Generated via simulation by the author using MATLAB (Mathworks Inc.); (*reconstructed image*) Mikael Häggström, CC0, via Wikimedia Commons. **MRI**: (*raw data*) Generated via simulation by the author using MATLAB; (*reconstructed image*) courtesy of Martijn Cloos and Graham Wiggins. **PET**: (*raw data*) Generated via simulation by the author using MATLAB; (*reconstructed image*) I. G. McKeith et al., ed. Lukelahood, CC BY 4.0, via Wikimedia Commons. **Ultrasound:** (*raw data*) Generated via simulation by the author using the MATLAB Ultrasound Toolbox (https://www.biomecardio.com/MUST/, D. Garcia, "SIMUS: An Open-Source Simulator for Medical Ultrasound Imaging. Part I: Theory and Examples," *Computer Methods and Programs in Biomedicine* 218 (2022):106726); (*reconstructed image*) © Nevit Dilmen, CC BY-SA 3.0, via Wikimedia Commons.

As intriguingly diverse as the raw data may be, the reconstructed images in figure 5.1 all take a consistent visual form. Let us zoom in on one of these images, in figure 5.2, to understand its organization. The image in question is a high-definition brain image obtained by MRI. The full image on the left is sufficiently detailed that you can appreciate the complex folding of the cerebral cortex, indicated by light gray curlicues against a darker gray background. Thin black lines crisscrossing the more central structures represent tiny blood vessels that literally supply food for thought, and carry away its products. When I look at an image like this, even after decades as an imager, I am awed by the intricate spatial organization of the organs that keep us alive and make us who we are. An image like this feels like a rabbit hole. It invites me to dive in, daring me to zoom all the way in

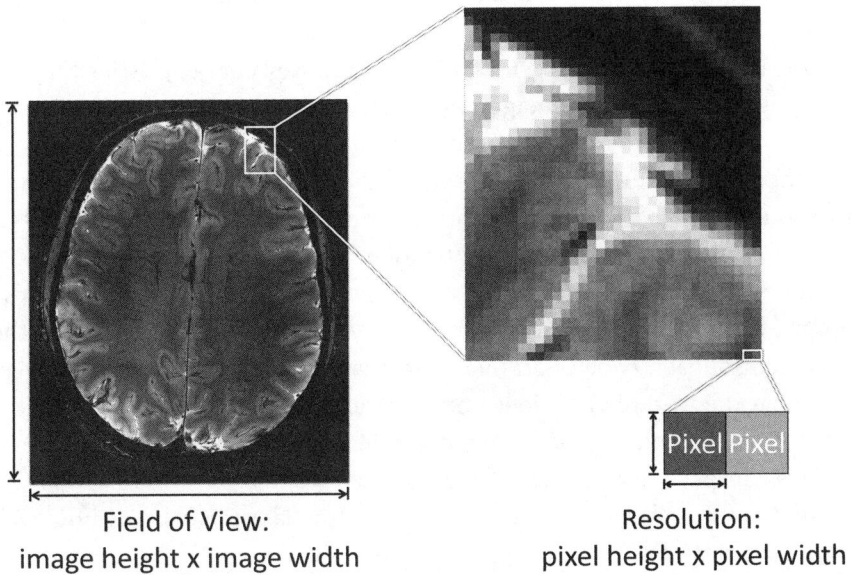

Field of View:
image height x image width

Resolution:
pixel height x pixel width

**FIGURE 5.2** The structure of a digital image. The field of view is the total extent of physical space represented in the image. Spatial resolution indicates the size of the smallest structures that can be distinguished when you zoom in. One approximate measure of resolution is the size of the smallest picture element, or pixel. Actual resolution depends on both the physics of image collection and the mathematics of image reconstruction.

*Source*: Brain MR image courtesy of Martijn Cloos and Graham Wiggins.

to the microscopic wonderland of neurons and synapses, of molecules and atoms that make up the living brain. When we do zoom in, however (on the right of figure 5.2), we see that the organization of the image itself is almost disappointingly simple. It is a regular rectangular grid.

Why choose such a simple organization to represent a complex structure like the brain? It makes sense that there are limits to the level of detail we can see—that there is a smallest possible picture element, like a silver grain in film. But in a computed image, unlike with film, we are free to choose the size and shape of this element. So why should we select an unnatural, sharp-edged rectangular shape rather than a curving shape more akin to brain structures? There are several reasons. One is that we do not know in advance where each bit of complex structure will fall in our image, and a regular grid at least treats all regions as equivalent. Another reason is that rectangles of constant size tile any flat surface in a consistent way that is easy to understand (and easy to reference, using grid coordinates like those on a map).[1]

The annotations in figure 5.2 indicate some key characteristics that tell you quite a lot about an image regardless of what it depicts. As a result, these characteristics tend to pop up repeatedly in scientific treatments of imaging. Casually dropping them into conversation is a great way to sound as if you know what you are talking about at cocktail parties populated by tech types. So grab your cocktail of choice, and let's get to it.

First is the "field of view." This is just a measure of the total extent of the depicted scene: how far you can see overall. If you ran a scale bar along the full extent of each edge of an image, this is how far the bars would reach. Note that the field of view does not represent the size of the image itself—that can be expanded or contracted at will. The field of view is the extent of physical space that the image represents.

Next is "spatial resolution." Technically, this is a measure of how well any two points can be distinguished from each other. For example, can you see two tiny blood vessels a millimeter apart, or are they blurred together? Spatial resolution is a means of quantifying the capacity for spatial *discrimination*—one of the two key features of any imaging approach, whether natural or artificial. While its precise definition can depend on what exactly you want to discriminate, in some simple cases that do not approach fundamental limits of detection (limits we will explore further in chapter 6), the spatial resolution of a human-made image may be quantified approximately as the dimensions of the smallest picture element in that image. This smallest picture element is called a pixel. It is a continuous region that shares a constant intensity or color, which does not change no matter how much you zoom in—like one of the gray squares on the bottom right of figure 5.2. As a modern reader, you are likely familiar with pixels from the specifications of your screens or cameras. A megapixel camera has the capacity to capture images with a million pixels each.[2]

Though we tend to view images head-on as upright boxes, it is worth remembering that the field of view of an image may extend along any direction in the space it represents and that we must orient ourselves accordingly to interpret its information content. Figure 5.3 illustrates some classic slice orientations used in medical imaging. As also shown in the figure, we can stack groups of adjacent 2-D images to form volumes. In this case, the field of view has a third dimension. We can also treat the image grid as extending to three dimensions, and we can replace pixels with "voxels,"

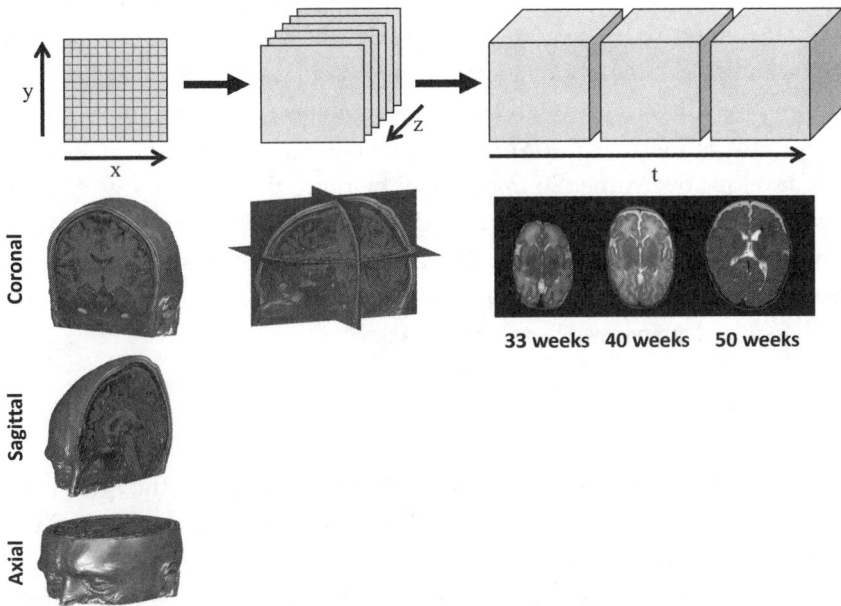

**FIGURE 5.3** Representing space and time in imaging. (*Left*) Images can be obtained in various orientations, such as the classic coronal, sagittal, and axial slices often used in medical imaging. (*Center*) Slices can be stacked into volumes, which we can then reslice as we wish. (*Right*) We can also obtain multiple imaging volumes over time, whether with dynamic imaging of moving tissues like the heart or with multiple imaging sessions separated in time, for example yielding images of a neonatal brain at various ages.

*Sources*: (*Bottom left*) Images generated by the author using MATLAB (The Mathworks Inc.) from an open-source, high-resolution brain MRI dataset described in F. Lusebrink et al., "T1-Weighted In Vivo Human Whole Brain MRI Dataset with an Ultrahigh Isotropic Resolution of 250 μm," *Scientific Data* 4 (2017): 170032; (*bottom right*) from J. Dubois et al., "MRI of the Neonatal Brain: A Review of Methodological Challenges and Neuroscientific Advances," *Journal of Magnetic Resonance Imaging* 53, no. 5 (2021): 1318–43, figure 8.

which represent the smallest volume elements in the grid. Whereas the orientation of an individual 2-D image is given by the particulars of the imaging process, once images have been assembled into a volume, we can slice through that volume as we wish using software, subject to limitations given only by the underlying spatial resolution and field of view. Three-dimensional imaging creates a virtual space that we can inspect as we see fit.

In fact, imaging does not stop at three dimensions. We can add as many dimensions to our grid as we like for the sake of convenient organization of information. For example, we can gather image slices or volumes dynamically, stacking them up according to the time at which they were gathered. If we want to see how a body looked in a particular orientation at a particular time, we can slice through the resulting dataset accordingly. As shown at the right of figure 5.3, for example, we can check in on the same brain as it develops over time. So images can be more than just representations of spatially organized information; they can actually convey volumes of information in space and time. Once the moment of data collection has passed and we have put our images in order, we can travel through our reconstructed spacetime however we wish.[3]

So much for the structure of images. Let us return now to the question of what information an image contains. From one perspective, the answer is simple: digital images contain numbers.

Each pixel in a modern digitized image has a distinct numerical value or set of values associated with it that characterizes its brightness or its color. Unlike natural images, in fact, digital images are stored in almost platonically abstract form as ordered arrays of numbers. I am reminded sometimes of the depictions of raw code underlying artificial realities in movies like *The Matrix*. Those inscrutable streams of numbers and symbols on-screen or in virtual space make for great cinema, but they are not far from the truth when it comes to imaging.

This means that the quality of images can also be captured in numbers. One particularly useful indicator of quality is the signal-to-noise ratio (SNR). SNR is a measure of sensitivity, the second key feature of any imaging approach. The signal in this ratio is what you want to see. The noise is everything else. In the case of ultrasound, the term *noise* is literal. Unlike the cleanly reflected ultrasound signal, noise is scattered sound that adds a nonsensical hum to the background and shows up as fuzz in the background of the image. For other imaging methods, *noise* is more metaphorical. The grainy background on a low-light photograph or a high-resolution microscope image looks messy, the way we expect noise to sound.

Where does this fuzzy, noisy background come from? It can come from anywhere and everywhere: the measurement apparatus recording light or projections, the object being imaged, the air, or even the cosmos itself.[4] The entire career of many imaging scientists, and in truth any scientists regardless of their specialty, can be framed as a pitched battle against noise. This is certainly true for me. It is also the case for my father, who, as a young particle physicist, had to pull faint subatomic signals out of the noise of subway trains passing underground beneath the laboratory. Rod Pettigrew—a character in our imaging story whom you will meet in chapter 7—let slip one of my all-time favorite scientific quotations at a workshop I once attended as a callow new imaging scientist. He said, "There are two things in life everyone wants more of. One of them is SNR." I once went through a period of such obsession with signal and noise, in my publications and research group meetings and day-to-day conversations, that my social worker spouse, Sarah, started dreaming about SNR. Psychologize that!

Leaving noise in the background for now, what information does the signal in an image convey? Here the answer is anything but simple. Image content is as diverse as the probes we use to make images. It is as varied as the transformations we have dreamed up to reconstruct those images. Figure 5.4 shows two noteworthy examples, one from outer space and one from inner space. The images in the top row show complementary views

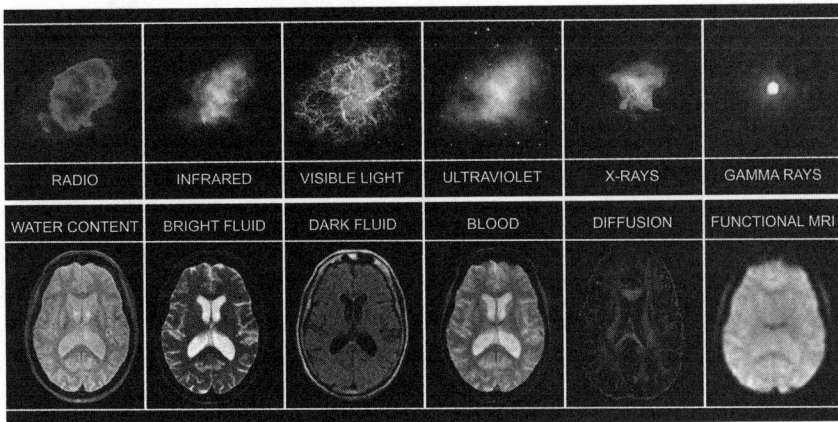

**FIGURE 5.4** Imaging shows us many sides of the same thing. (*Top*) Different views of the Crab Nebula from telescopes sensitive to different parts of the electromagnetic spectrum. (*Bottom*) Different representations of the same human brain using various flavors of MRI.

*Sources*: (*Top*) Buehler, CC BY-SA 3.0, via Wikimedia Commons. (*Bottom*) Based on S. Duchesne et al., "Structural and Functional Multiplatform MRI Series of a Single Human Volunteer Over More Than Fifteen Years," *Nature Scientific Data* 6 (2019): 245, figure 1.

of the Crab Nebula captured by telescopes sensitive to different parts of the electromagnetic spectrum. The bottom row shows brain images of the same person obtained in a single imaging session using different flavors of MRI. Every image in the figure has a story behind it: how it was made, who made it, and what it meant to them. I will tell you a few of those stories now so you can begin to appreciate just how many different ways it is now possible to see one and the same thing.

The images at the top of the figure convey two stories at once: a modern story of persistent human inventiveness, and an ancient story of stellar Armageddon. According to NASA, the Crab Nebula is "the shattered remnant of a massive star that ended its life in a supernova explosion."[5] The nebula measures five or more light-years end to end and is 6,500 light-years distant from the little blue marble on which we live. We know all this from a long history of observation, starting in 1054 CE when Chinese astronomers first documented the appearance of a new star in the constellation of Taurus. Of course, it had taken light from the actual explosion 6,500 years to reach those astronomers' eyes, and, when it finally did so, it found them in the era before telescopes. Since then, generations upon generations of astronomers have trained their progressively enhanced vision on the nebula.

The radio wave image in figure 5.4 comes from the Karl G. Jansky Very Large Array (VLA), a group of twenty-eight towering telescope dishes on the Plains of San Agustin in central New Mexico. The VLA's dishes perform coordinated measurements of radio waves beaming toward us from space, and these measurements are converted into images whose pixels represent the strength of the incoming waves from any given part of the sky. Radio waves are on the low-energy end of the electromagnetic spectrum, and radio astronomy, a field established in the 1930s, concerns itself with emissions from comparatively low-energy cosmic processes. In the case of the Crab Nebula, those emissions are believed to originate in a pervasive wind of particles driven by a pulsar at the nebula's core.

While radio astronomy may tell us about low-energy effects of the pulsar at the heart of the Crab Nebula, the Crab Pulsar itself is anything but low energy. It is the collapsed core of the supernova, and its gravity is strong enough to mash the normal components of atomic nuclei together to create a slurry of uncharged particles called neutrons along with layers of charged particles and other exotic matter. In other words, it is a neutron star, a strange and powerful beast which is about as close as something can be to a black hole without actually being one. To make matters more dramatic, this particular neutron star, which is the size of Manhattan, is spinning in space at the mind-boggling rate of thirty times a second. As it spins, it drags with it a strong magnetic field—ten million times as strong as our strongest MRI magnets—that sweeps through the surrounding cloud of cast-off matter from the supernova. When charged particles like electrons encounter the sweeping field of the pulsar, they spiral around it and generate electromagnetic radiation, which shoots out into the universe like a gyrating, supercharged lighthouse beam. The low-energy radio wave emissions captured by the VLA are believed to originate from calmer regions of this post-stellar dynamo. Higher-energy emissions have other things to tell us about the Crab Nebula's dynamics.

Let's climb one step up in energy to the infrared. Infrared astronomy dates back to the 1830s, some thirty years after infrared light was first discovered.[6] It had a renaissance in the mid-twentieth century thanks to various technical developments, including the ability to launch telescopes into space. Infrared telescopes are generally similar to visible-light telescopes because rays in the near-infrared region of the spectrum are close enough to visible light rays that they can be bent by similar optical structures. However, many infrared frequencies are absorbed by water vapor in the earth's atmosphere—hence the value of orbiting telescopes, which can do their observing outside the haze of the atmosphere. The Spitzer Space Telescope captured the infrared image shown in figure 5.4, which also highlights clouds of energetic electrons trapped in the magnetic field of the pulsar. Note, though, that the distribution and relative intensity of structures highlighted in the infrared image differ from what can be seen in the radio image.

The Crab Nebula got its name from the crab-shaped appearance of images observed in mid-nineteenth-century visible-light telescopes. The Hubble Space Telescope, launched in the last decade of the twentieth century, upped the visible-light ante. The inner core of the Hubble visible-light image in figure 5.4 shows effects of the pulsar's magnetic field that extend into the visible range of the spectrum. However, the fine filaments on the outer fringes are believed to indicate the remains of the original exploding star's gaseous material as it continues to blast outward, generating light as it goes.

Seeing anything at all in the ultraviolet range required a telescope launched by spacecraft, since the earth's ozone layer filters out most ultraviolet rays. Therefore, the field of ultraviolet astronomy didn't hit its stride until the 1960s. Decades later, the ultraviolet telescope aboard NASA's Swift satellite would capture the ultraviolet image in figure 5.4. This image provides yet another complementary view of the combined effects of the supernova blast and the pulsar lighthouse, at even higher energies.

The Chandra X-ray Observatory, which also sits safely beyond Earth's atmosphere, went further still. The Crab Pulsar dynamo is powerful enough to generate x-rays by accelerating charged particles to near the speed of light. X-rays, as you know from chapter 3, have enough energy to blast through ordinary objects and give us the projections we use here on earth for medical imaging among other things. This means that ordinary mirrors and lenses won't be very effective at bending and collecting x-rays. Chandra's mirrors, therefore, had to be constructed from special dense materials. They also had to be polished to exquisite smoothness and configured not for traditional reflections but for glancing bounces that would bring the high-energy beams into Chandra's x-ray detectors. The result of all this careful engineering is images like the x-ray image shown in figure 5.4. You can see that some of the outer blush in the lower-energy images is gone, leaving us to focus on the higher-energy interior. The bright central dot betrays the location of the pulsar itself. Jets of matter and antimatter emerge from the poles of the pulsar. You can also see rings that are thought to mark a shock wave where the pulsar-driven material meets the rest of the nebula.

Even more intense bursts are thought to result in high-energy gamma-ray emissions. Gamma rays are at the far high-energy end of the electromagnetic spectrum. They are the beams that emerge when electrons meet anti-electrons in PET imaging. As it happens, the process can also run in reverse, with gamma rays producing electron–positron pairs. This is how devices like NASA's Fermi Gamma-ray Space Telescope detect incoming gamma rays. Bending gamma rays is an even more difficult proposition than bending x-rays, though physicists are working on ways to do it. In the meantime, spatial maps of gamma-ray emission can be built up by back-tracking the angle at which each gamma ray comes into specialized layered detectors. This process results in images like the one shown at the far right of figure 5.4. While the Crab Nebula's other emissions tend to be as reliable as clockwork (so much so that the nebula is sometimes used as a calibration standard for brightness and timing in some areas of astronomy), its gamma-ray emissions have been observed to flare up unexpectedly.

These flares are believed to occur when the pulsar's magnetic field shifts configuration suddenly. The gamma-ray image in the figure, captured by Fermi, shows a snapshot of such a flare, which you can see emerging from the vicinity of the pulsar. The Crab Nebula is now known as one of the most intense sources of gamma rays in the observable universe.

As you can appreciate from this eclectic set of images, each small step in our capacity to see and record complementary spatial information has resulted in giant leaps forward in our understanding of the strange entities that populate our cosmic neighborhood. With just one exception, all the images in the top row of figure 5.4 record frequencies of light that are not naturally visible to us, converted into colors we can see. In fact, while the images here are shown in grayscale, different features can be assigned different colors at will, and different frequencies can even be combined together in multicolor composites that convey various types of information at once. Such a composite image is shown in the color insert (plate 5). Images like this perform a special visual alchemy. They make invisible signals visible, presenting to our limited eyes some evidence of the universe's wider design.[7]

The stories told by the images in the bottom row of figure 5.4 take place much closer to home. These images document the health of one person's brain. They originally appeared in a scientific article describing hundreds of MR images of a single volunteer acquired in more than seventy sessions using thirty-six different MRI machines over the course of more than fifteen years.[8] The set of six images in the figure followed the Canadian Dementia Imaging Protocol, designed to explore a wide range of brain health issues related to aging. All the images share the same orientation and field of view. Each, however, has a different appearance, which results from giving the MRI scanner a slightly different set of operating instructions each time. In some images, fluid in the brain is bright; in others, it is dark. Particular brain tissues are highlighted to varying degrees in the various images. The difference in appearance of distinct tissues is known in the imaging field as contrast. Among medical imaging modalities, MRI is the king of contrasts—so much so that extensive training is required to sort out what all the available types of contrast mean. But MRI doesn't see across the whole electromagnetic spectrum—it just uses radio waves. So, how are all these different image appearances possible?

That story begins back in the 1970s with Raymond Damadian and Paul Lauterbur. In describing the origins of MRI earlier, I mentioned that the faint radio signal emitted by water in the body is transient. In fact, the magnetic resonance signal often decays away in a matter of seconds or fractions of seconds. This means that the strength of the signal you measure depends on how long you take to measure it. Moreover, the rate of signal decay is not the same in all tissues. Signal from one tissue may decay faster than signal from another tissue, owing to subtle differences in the microscopic chemical and magnetic environment occupied by water in each tissue. This means that, after some time, the signals from equal quantities of two different tissues will no longer be equal, and you will get a distorted view of the relative importance of those tissues.

Damadian recognized that this property represented a remarkable opportunity for imaging—an opportunity to highlight some tissues more than others. His 1971 *Science* paper showed that tumors could be detected in tissue samples by carefully timed magnetic resonance experiments, based on the different rates of signal decay in tumors as compared with normal tissues.[9] It was this observation that set him on the path of trying to create images with magnetic resonance. He was convinced that such images would be able to pick out tumors buried within normal body structures. Unlike his Nobel ambitions, this conviction would be vindicated in dramatic fashion by the subsequent history of MRI.

Lauterbur, for his part, also appreciated the potential for diverse contrasts in MRI. The third illustration in his landmark 1973 *Nature* paper showed that different rates of signal decay could result in striking variations in image brightness. The illustration also provided a striking reminder that the absence of signal can sometimes be as informative as its presence. Just as blocking light can give information in the form of shadow pictures and tracking the attenuation of x-rays can produce informative projections, noting changes in signal strength under various conditions can invest MRI projections with a wide range of information. Lauterbur went on to predict, correctly, that many types of image contrast would be possible with MRI, based on chemical composition, the rate of water diffusion, and other properties of objects or tissues.[10]

It took until the 1980s for MRI to become fully practical in human subjects. 1986 turned out to be a banner year for image contrast in MRI. In that year, two key papers, both from German research groups, appeared in the journal *Magnetic Resonance in Medicine* (whose name already asserted, even in those early days, that magnetic resonance had a role to play in medicine). The first paper, by Jens Frahm and colleagues, introduced the so-called FLASH technique, which was fast enough to visualize the beating heart.[11] The second paper, by Jürgen Hennig and colleagues, described the RARE rapid imaging technique.[12] Both techniques worked by hitting the body with radio wave pulses again and again in rapid succession. The two methods resulted in images with strikingly different contrast, however. With RARE, enough time passed between pulses that only the tissues with slowly decaying signal remained bright by the end of the sequence. With FLASH, on the other hand, the barrage of pulses was so rapid-fire that it beat down persistent slow-decaying signal until it was smaller than the fast-decaying signal, and it was the tissues with fast-decaying signal that paradoxically appeared bright.[13] Details aside, what Frahm and Hennig had devised were ways of making the same tissues take on radically different appearances, all within imaging times short enough for practical use in patients. Complementary approaches or "pulse sequences" like these, providing complementary information about tissue composition, would become the basis for routine image contrast in MRI protocols like the one shown in figure 5.4. The first four MR images in the figure were all generated with contrast based on differential signal decay but, as you can see, they have strikingly different appearances. The leftmost image minimizes the effects of signal decay, showing the raw water content of tissues. The "bright fluid" and "dark fluid" images were obtained by manipulating signal decays in a way that makes cerebrospinal fluid brighter or darker, respectively, than other brain tissues. The visibility of various classes of abnormalities, including tumors, has been shown to be selectively enhanced in one or another of these image types. Similarly, the pulse sequence used to generate the image labeled "blood" was designed to highlight the signal-decay effects of excess blood that would appear in a brain hemorrhage. Just as the various Crab Nebula images offer complementary views of the nebula's dynamics, the various MR image contrasts provide complementary windows into the health of living tissues. From these examples it is clear that in medical imaging, just as in astronomy, you really need to know how an image was constructed in order to understand what it is telling you.

Returning to 1986, contrast based on water diffusion also made its appearance that year, delivering on Lauterbur's prescient prediction. Rather than just letting signal decay take its natural course, Denis Le Bihan and his coauthors showed that it could be sped up in informative ways. Their technique, first published in the journal *Radiology*, set up conditions in which fast-moving water molecules would lose signal faster than slow-moving ones.[14] As no less a figure than Einstein had noted in his investigations of Brownian motion (the random jittering of small particles immersed in water, as seen in microscopes since the nineteenth century), diffusing water molecules bump their way around on a random walk. Le Bihan knew that when this random walk takes place in a magnetic field gradient, the resulting magnetic resonance signal is also randomized, and the rate of signal decay depends on the rate of diffusion. This means that the brightness of an image could be made to depend on the rate of diffusion. Why is this useful? It just so happens that the distance typically traveled by a diffusing water molecule during each little segment of an MRI pulse sequence is on the same order as the dimensions of a typical cell. If enough water molecules bump into cell walls or other microscopic structures along their way, these structures will slow them down, and we will observe changes in the diffusion rate. Therefore, even though each MRI voxel is generally millions of times larger than an individual cell, diffusion contrast can tell us a lot about what is going on at the cellular level. By using diffusing water as a probe, we can imprint microscopic cellular information on our macroscopic images.[15] This microstructural diffusion information was later proven to be useful in detecting strokes, cancer, and other disorders. Imaging scientists would also realize that characterizing differences in diffusion rate along different directions would allow them to trace out the complicated paths of nerve fibers. The "diffusion" image in figure 5.4 is a simplified depiction of directional diffusion in the brain.

By 1990, so many different types of biophysical and biochemical information had been loaded into MR images that the October issue of *Science* magazine sported a brain image on its cover and featured an article summarizing advances in "functional magnetic resonance imaging in medicine and physiology."[16] Another *Science* cover only a year later, in November 1991, heralded a new kind of "functional MRI" that offered a revealing new window into the functioning of the brain.[17] (You can find both covers reproduced in plate 4 in the color insert.) This new functional MRI, which has come to be known as "fMRI," works by tracking yet another microscopic source of signal decay: the difference in signal decay rates between blood that is fully loaded with oxygen and blood that has been depleted of oxygen. This difference was first exploited to create image contrast in 1990 by the Japanese scientist Seiji Ogawa, who was working at AT&T's Bell Laboratories at the time.[18] Electrical activity in the brain increases both the delivery of blood and the extraction of oxygen from blood, and it was quickly realized by independent research groups at the Massachusetts General Hospital, the University of Minnesota, and the University of Wisconsin that Ogawa's new oxygen-dependent contrast mechanism could be exploited to observe which parts of the brain are active at any given time.[19] These teams showed that after suitable transformations of the underlying signals, brain activity shows up as bright spots on images generated with fMRI. Around the world, a frenzy of exploration followed, and it continues to this day. Imaging researchers have figured out how to separate the baseline activity of the brain at rest (as shown in the "functional MRI" image in figure 5.4) from activity associated with particular tasks like tapping your fingers or watching a movie. Scientists of many different stripes use fMRI to tease out how our brains work. Functional MRI has also captured the popular imagination, just as photographs and x-rays and space-borne cameras and early tomographic slices did earlier in the history of imaging. Functional MRI, though, has an almost literal hold on our imaginations. After all, it shows us our brains in the process of thinking.

In explaining the rudiments of what distinguishes the images in figure 5.4, I've barely scratched the surface of MRI contrast, but I think you get the idea. The information that images can convey is limited only by the laws of physics and the boundaries of human ingenuity. With modern MRI, you can visualize streamlines of flowing blood, as you can with Doppler ultrasound. Both ultrasound and MRI can be used to probe the rigidity of tissue, in direct analogy to our evolved sense of touch. With suitable pulse sequences, MRI can be made sensitive to specific molecular signatures, like a kind of artificial sense of taste. As I mentioned back in chapter 4, designer contrast agents allow us to sniff out particular molecules using PET and other nuclear imaging methods. Various nonradioactive contrast agents can also be injected or ingested during MRI, CT, ultrasound, and optical imaging studies, allowing imagers to highlight areas of the body where these snooping agents end up. Even the images emerging from ordinary cameras nowadays are no longer what-you-see-is-what-you-get. They can be color coded to show heat signatures, or filtered to highlight edges, or transformed to simulate the view through, say, a butterfly's compound eye. A modern image is like a secret message, waiting to be decoded. All you need is the right key.

# 6

# Pushing the Limits

*How the quest to extend the reach of our vision has continued.*

N ow that you know how a modern image is constructed, and what kinds of information it can contain, it's time to tackle a question we haven't addressed so far: what are the limits of artificial vision? What principles govern how much we can magnify a scene or how much detail we can capture? These principles are like the unfilled edges of a map. They demarcate the current limits of our knowledge. They also serve to motivate bold explorers drawn to terra incognita.

In chapter 2, you learned about the pioneering efforts of early astronomers and microscopists, and subsequent chapters have cataloged a rich set of technologies and transformations that let humans see in myriad new ways. In this chapter, we will consider scientists' valiant struggle in recent times to push the limits with microscopes, telescopes, and other imaging devices. One might be tempted to view their story as a story of technological refinement, but that, I think, would be a mistake. Just as their distant imaging ancestors were, these modern pioneers are driven by a primordial urge to see farther, and more clearly. They continue to push frontiers not through the physical distances they travel but through the tools they build to bring other worlds to us.

It may not surprise you that the limits pushed by imagers from time immemorial relate to the two fundamental determinants of imaging performance: spatial discrimination and sensitivity. In this chapter, we will begin with advances in the resolving power of microscopes. We will then consider analogous developments in the design of telescopes. Finally, we will circle back to medical imaging to explore sometimes hidden connections among the various modern disciplines of imaging.

Physics has long been concerned with finding the smallest constituents of things: molecules, atoms, hadrons, quarks, strings. This obsession with zooming in has engendered a very practical question that has driven generations of imagers: what is the smallest thing that we can see?

This question brings us back to optics. The designers of early microscopes quickly learned to maximize magnification by manipulating the shape and spacing of lenses. They also had to concern themselves with the precision manufacturing of smooth and symmetrical lenses, as well as stable microscope housings, to make sure images weren't distorted by imperfections or blurred by motion. Eventually, when these details were more or less under control, other limits were uncovered. In 1873, the German physicist Ernst Abbe posited that, even for perfect lenses, the sharpness of optical images would be limited via a relation that came to be known as the diffraction limit.

As you now know, the smallest details observable in an image are dictated by the spatial resolution of the image. As I hinted at in chapter 5, true resolution is not always, or even usually, equal to the nominal pixel spacing in an image. Instead, resolution characterizes our ability to distinguish two nearby points in the scene being imaged. What might prevent us from distinguishing two separated points unambiguously? Well, optical imaging relies on tracking lines of sight for bent light. Light, however, is not made up of straight, idealized rays. It is an electromagnetic wave, and as such it has an annoying tendency to spread out in a process called diffraction.

Diffraction occurs whenever light passes through a narrow opening or aperture. The extent of the spreading depends on the width of the aperture via an uncertainty principle: the smaller the aperture, the bigger the spread (figure 6.1, top). This principle holds true for all waves.[1] A water wave on the open ocean may go on its merry way in more or less a straight line, but when it hits a narrow constriction, like a break in a long dock, it spreads out in concentric circles from that point onward. The same is true when light passes through a lens (figure 6.1, bottom). If the lens opening

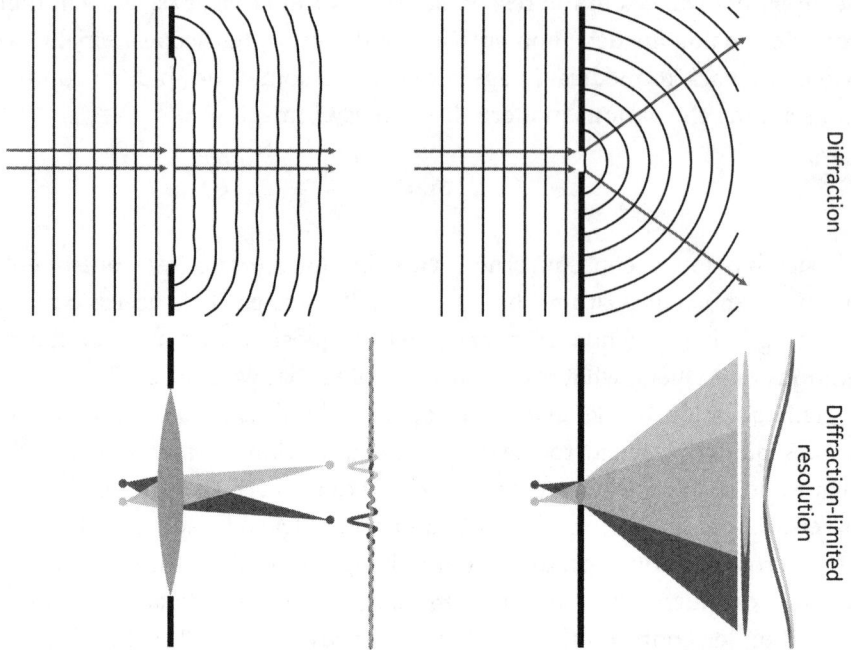

**FIGURE 6.1** The diffraction limit. (*Top*) Light travels as a wave with repeating peaks and troughs. When a wave passes through a wide opening (*left*), it ripples a little at the edges but proceeds largely unperturbed in the middle, whereas a narrow opening (*right*) causes it to spread out widely through a process called diffraction. Thin black lines indicate wave peaks. The separation between peaks is the wavelength. Gray arrows are light rays, which indicate the direction of travel of the wave and which are used in ray-tracing diagrams, as shown in chapter 2. (*Bottom*) Spreading due to diffraction can counteract the focusing effect of lenses. Light emerging from two nearby points is focused to distinct points when passing through a wide lens (*left*), but when the size of the lens opening is close to the wavelength, light spreads out, and the points become indistinguishable (*right*). Differently shaded wedges indicate light rays emerging from each point and passing through the lens. Shaded curves to the right of each panel indicate the individual diffraction patterns of light arriving at a screen or detector array.

*Source*: Daniel K. Sodickson.

is too small, the light spreads out, counteracting the focusing effect of the lens. If that light is on its way to an array of optical detectors (like a retina or a piece of film or an electronic recording device), its spreading means that the signal that would ideally have been recorded in one detector can leak into other adjacent detectors, making the edges of the image fuzzy.

Effectively, there are multiple lines of sight for each spreading light beam rather than just one, adding a blur of uncertainty to diffraction-limited images. Each point in the object no longer appears as a single point in the image but is instead associated with an extended wavy diffraction pattern. In the two-point example shown at the bottom of figure 6.1, the wide opening results in individual diffraction patterns that are narrow and well resolved. For the narrow opening, the two diffraction patterns are broad and highly overlapping, and the points become nearly impossible to distinguish.

Abbe's diffraction limit is simple to state. The minimum distance between two points that can reliably be distinguished in an optical image is given by one-half the wavelength of light divided by a quantity known as the numerical aperture, which measures the size of the opening through which light is admitted. For green light of 500-nanometer wavelength and a state-of-the-art numerical aperture, the minimum resolvable distance can be as small as about 150 nanometers, which is smaller than most biological cells but larger than most viruses, and quite a bit larger than the majority of protein molecules. The diffraction limit constitutes a clear line in the sand for developers of optical imaging devices. Any attempt to zoom in further than this limit would in principle represent wasted effort, since magnified features would just be blurred out again by diffraction.

As I said, the diffraction limit is a fundamental property of waves. Enterprising inventors in the twentieth and twenty-first centuries, though, found a surprising number of ways around this supposedly fundamental limit.

One simple way to improve diffraction-limited resolution is to widen the opening for microscope lenses, increasing the numerical aperture and reducing the spreading of incoming light. For microscopy, the benefits of wide apertures are not without their own limits, however. At some point, even a lens of modest size captures just about all the light emerging from a microscopic object, as long as it is sufficiently nearby. There is then no benefit to widening the lens opening further.[2] One other obvious approach to resolution enhancement is to use a smaller wavelength of light, making the minimum separation of distinguishable points correspondingly smaller. For this, one needs higher-energy light than can be found in the visible part of the spectrum. Since Röntgen, we have known that high-energy light can be harnessed in the form of x-rays. But using x-rays directly for traditional magnification is challenging because powerful x-rays are difficult to bend. Eventually, x-ray microscopy became possible, leveraging advanced materials and high-intensity x-rays beams, such as those spun off the accelerators physicists use to study elementary particles.

Long before people figured out practical x-ray optics, though, they devised another clever imaging technique that allowed them, in special cases, to resolve features all the way down at the atomic scale: around 0.1 nanometers, well below Abbe's diffraction limit even for today's best x-ray microscopes. This technique actually embraced diffraction as an asset rather than a liability. When x-rays are directed onto a regular lattice of atoms in a crystal, and when the wavelength of those x-rays is on the same order as the spacing between atoms, the high-energy x-ray waves spread out and create an interference pattern, like ripples crossing one another in a pool. If the crystal or the x-ray beam is rotated to generate different interference patterns, then mathematical transformations can be used to sort out what particular arrangements of atoms must have produced the resulting patterns. In fact, deducing the internal structure of a crystal from distinct x-ray diffraction patterns (which may be viewed as complicated projections of that structure created by passing x-ray waves) might now be recognized as a form of tomography, though the development of x-ray crystallography predates even Radon's 1917 treatise on reconstruction from projections. Two successive Nobel Prizes went to the physicists who established the foundations of x-ray crystallography: Max von Laue in 1914 and the father–son team of William Henry Bragg and William Lawrence Bragg the following year. More than ten subsequent Nobels have been awarded for the discovery of important molecular structures using x-ray crystallography. This list includes the famous structure of the DNA double helix, for which three men—James Watson, Francis Crick, and Maurice Wilkins—received Nobel recognition and one woman—Rosalind Franklin—notably did not.

Strictly speaking, another early advance in high-resolution microscopy worked within the bounds given by Abbe, but with a twist: it didn't actually use light. As I mentioned earlier in this chapter, the diffraction limit applies to all waves, and for the best resolution one wants to use waves with the smallest possible wavelength. What has a very small wavelength but can still be bent to produce magnification? According to quantum physicists, an electron does. In his 1924 doctoral thesis, Louis de Broglie postulated that particles can behave like waves and that the more massive a particle is, and the faster it moves, the smaller its wavelength becomes.[3] The diffraction-limited resolution associated with a beam of particles such as electrons can be quite high. Conveniently, electrons are also charged particles that can be steered using electric and magnetic fields. The necessary ingredients, then, are all there: bend beams of electrons in place of light beams, and you have yourself an electron microscope (figure 6.2, left).

Given the history of aggressive scientific striving we've traversed so far, it may not surprise you to learn that the question of who invented the electron microscope is a subject of some controversy. Suffice it to say that the first working devices appeared in the 1930s, and they went on to set new records for magnification. A Nobel Prize in Physics went to Ernst Ruska in 1986 for his seminal contributions to electron microscopy. Today, we owe our understanding of the spiky shape of the COVID-19 virus to electron microscopes. Meanwhile, another form of electron microscopy allowed us to zoom in on the structure of individual viral spikes. The Nobel for that technique, called cryo-electron microscopy, was awarded in chemistry in 2017 to scientists who had figured out how to map the three-dimensional structures of frozen molecules atom by atom using various projections with electron microscopes. Yes, I'm talking about tomography with electrons![4] Cryo-electron microscopy circumvented the diffraction limit for electrons, much as x-ray crystallography had for x-rays, by relying on reconstructions from projections.

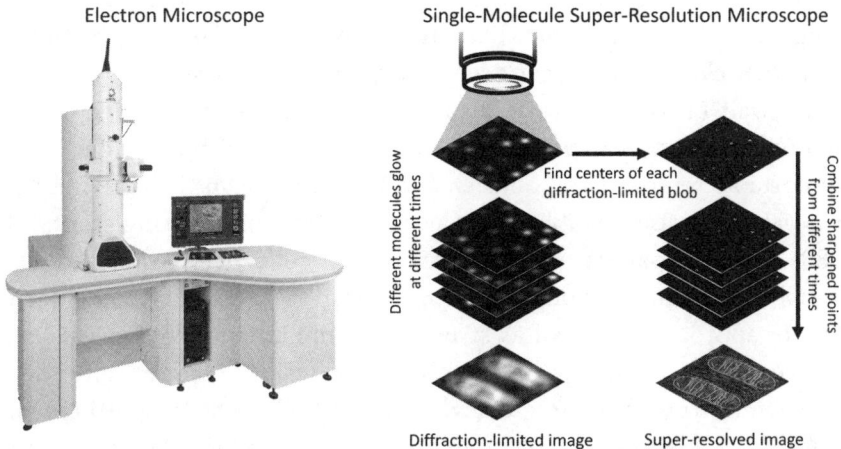

Electron Microscope     Single-Molecule Super-Resolution Microscope

Different molecules glow at different times

Find centers of each diffraction-limited blob

Combine sharpened points from different times

Diffraction-limited image     Super-resolved image

FIGURE 6.2 Pushing the limits. (*Left*) A modern electron microscope. (*Right*) Principles of operation of single-molecule super-resolution microscopy.

*Sources*: (*Left*) Courtesy of JEOL Ltd.; (*right*) generated by the author using MATLAB (The Mathworks Inc.), starting with an image of mitochondria (Barou abdennaser/Shutterstock).

Sharing the Nobel with Ruska back in 1986 were two physicists, Gerd Binnig and Heinrich Rohrer, who, just five years or so earlier, had developed yet another limit-busting variant of electron microscope called a scanning tunneling electron microscope. In this device, electrons were used not for their short wavelength but for their ability to flow (or "tunnel") in a controlled way between a sharp needle and a surface over which the needle was suspended. By scanning the needle back and forth across the surface, with the needle kept at a fixed offset regulated by the flow of electrons, it was possible to map out the surface contours of tiny objects. The idea is a little like running your finger over a bumpy surface to feel where it rises and falls. Now imagine that your finger is just a single atom thick. Score another win for seeing small things, in this case using other small things as probes, and emulating our evolved sense of touch.[5]

Believe it or not, the very same approach turned out to work for light as well. In a photon scanning and tunneling microscope, photons do the tunneling in place of electrons as a tiny tip is scanned along a surface. Other forms of high-touch optical microscopy became possible once light sources could be moved sufficiently close to the objects being imaged. Within a few wavelengths' distance from a source, before its full wave nature comes together, light can behave transiently like a rapidly decaying electromagnetic field, which can react to the fine structure of nearby surfaces. In other words, you can beat the diffraction limit by sidling up so close to something that light no longer acts like a traditional wave. When you've set your sights that close, you just can't see very far beneath the surface.

But wait (as TV ads for Ginsu knives and other mail-order marvels announced when I was young). There's more. Even with light sources at a comfortable remove from objects, and even when light is behaving itself as a wave, it is still possible to get around diffraction. Starting around the 1980s, a whole parade of techniques, which collectively would come to be called super-resolution microscopy, began to appear. They included a class of approaches known as structured illumination, which used finely patterned light to highlight the fine structure of small objects. Some methods even borrowed the frequency-encoding trick of MRI, splitting white light into its component colors and directing different colors at different parts of an object so that resolution was determined by the spatial separation of distinct frequencies rather than by focusing optics. The panoply of other themes and variations, not to mention their associated technical acronyms, can be dizzying to sort out.

Here, I will call your attention to just two particularly influential classes of super-resolution optical microscopy methods. The first uses two distinct and carefully tailored light sources, one stimulating fluorescent molecules to glow, and the other canceling out fluorescence from all but a small non-diffraction-limited spot. The resulting glow can then be localized precisely to the spot in question, and that spot can be scanned across an object to build up an image. Stefan Hell got the Nobel nod in chemistry in 2014 for this stimulated emission depletion (STED) approach, which was billed as a new "nanoscopy."

Also on the Nobel stage with Professor Hell that year were Eric Betzig and William Moerner, who had devised something called single-molecule microscopy. Their nanoscopic method also relied on light emerging from small non-diffraction-limited spots—in this case individual molecules whose fluorescence can be turned on and off through light-activated or chemically activated random processes. In essence, different random subsets of molecules in an object to be imaged are made to glow at different times. As a result of the diffraction limit, the size of each glowing blob in an image taken at a particular point in time will be much larger than the size of the corresponding molecule. However, when the number of molecules glowing at any given time is sufficiently small, the average spacing between the blobs will be larger than the size of each blob, and the individual blobs can be distinguished easily. Each individual blob can then be narrowed down artificially to a molecule-sized spot at its center since one can be confident that only one molecule is contributing to each separate blob. The result is an image made up of sparse but narrow spots. Additional images of other random molecular fluorescences can then be gathered over time and sharpened up in the same way. A composite image collecting the narrowed spots from all these component images will then constitute a dense molecule-by-molecule map of structures smaller than the diffraction limit (see figure 6.2, right). Picture a detailed pointillist painting assembled by a team of artists dabbing impulsively with thick brushes at various parts of a canvas, and a companion team of fastidious scribes carefully placing a small dot of ink at the center of each paint blob, then wiping away the original paint. Chaplinesque comical overtones aside, such a complex process does get the job done, even if it takes more time and effort than a traditional snapshot.

Note that both nanoscopy methods I've just surveyed require recording very small quantities of light emerging from a tiny spot or even a single molecule at a time. Reflecting on how this feat can be accomplished brings us to another question of limits—in this case, of sensitivity. What, we now must ask, is the smallest optical signal we can observe?

Quantum mechanics once again gives us an answer. The smallest unit of light is, of course, a single photon. The challenge for designers of artificial photodetectors, then, was to engineer reliable detection down to the single-photon level, like the dark-adjusted eye. By now, modern photodetectors have comfortably surpassed our eyes in terms of raw sensitivity, having higher detection efficiencies over a wider range of frequencies. Chemical means of capturing light, like film, have largely been supplanted by electronic devices, such as photomultiplier tubes and charge-coupled devices (CCDs). Large arrays of small photodetectors in CCD arrays can now stand in for our retinas, gathering light from multiple positions in parallel.

Then there is the question of how to deliver enough light to a small object to yield detectable reflections, or to cause molecules in the object to respond by emitting requisite numbers of fluorescent photons. Specialized light sources represent another key element in the toolkit of modern microscopy. Lasers, first developed in the late 1950s and early 1960s, are used routinely to provide strong, coherent light at well-defined frequencies.[6] Illumination can be delivered broadly to an entire sample (widefield microscopy), passed through a pinhole to illuminate one spot at a time (confocal microscopy), or delivered from the side to get an entire surface glowing at once (light-sheet fluorescence microscopy). A veritable Lego kit of modular parts is now available to be assembled on vibration-damped optical tables in laboratories or packaged into sleek desktop assemblies by microscope companies.

Meanwhile, scientists have even figured out how to engineer the bodies being imaged. Chemists have designed all manner of custom fluorescent dyes—like those used in super-resolution microscopy—that glow in color when exposed to particular frequencies of light. These tailored molecular reporters hark back to opsins in the eye. They also take inspiration from some of nature's other colorful innovations. Engineered variants of the green fluorescent protein found in bioluminescent jellyfish were deemed worthy of a chemistry Nobel in 2008. Tiny glowing semiconductors called quantum dots followed suit in 2023. Biologists have taken up the baton and engineered experimental animals to express fluorescent proteins for easier imaging. (Cages full of glowing green mice. Seriously.) They have developed biochemical clearing techniques to make tissues in some organisms transparent so that light can probe otherwise hidden depths. A particularly ingenious cheat on resolution limits is called expansion microscopy, in which tissues are made to swell so that the same underlying structures can be seen in more detail at any given spatial resolution. If you can't improve your microscope, the reasoning goes, then just make the thing you're imaging bigger. Take that, diffraction! Alas, such an operation will not work, at least at present, with either human or heavenly bodies.

The twenty-first century has been a revolutionary time for optics, so much so that the age-old science of manipulating light has been rebranded with the new buzzword *photonics*. Physicists and chemists, conducting a quantum symphony of interacting photons and atoms, can probe the structure of complex materials comfortably below the diffraction limit. Biologists now have at their disposal a huge arsenal of tools for imaging life in real time. They can zoom in on the microstructure of living tissue like a web surfer using a fine-grained version of Google Maps. They can compare the distinctive views delivered by diverse photonic probes dutifully reporting in at various wavelengths. They can see the same tiny things in many different ways. In this juggling of complementary views, modern microscopists have quite a bit in common with their colleagues at the opposite end of the size scale. So, let us now check in on modern astronomers to see how they have dealt with the limits of resolution and sensitivity in their imaging instruments.

Like microscopists, astronomers seek to see ever-smaller things. The objects of astronomical inspection, however, appear small only by virtue of being very far away. This means that, unlike microscopists, astronomers cannot control the conditions around their imaging targets. Astronomical observers are relegated to passive observation from an insurmountable distance.

If one seeks to see things that are farther and farther away, one must find means of increasing both spatial resolution and sensitivity. Spatial resolution in astronomy is determined by the minimum distinguishable separation of points in the sky. This is generally characterized by differences in pointing angle rather than by linear separation since two objects showing up next to each other in the sky—say, a planet in our solar system and a gas cloud in another galaxy—may actually sit at dramatically different distances from us here on Earth. The diffraction limit on angular resolution for telescopes is quite similar to the limit formulated by Abbe, with the minimum angular separation essentially given by the wavelength of light divided by the diameter of the aperture of any light-collecting lens or mirror.[7] The sensitivity limit, meanwhile, is given by the smallest quantities of light that can reliably be distinguished from noise. Even bright objects dim with distance since all but the narrow cone of light that happens to be directed right at an observing telescope ends up lost in space, and the observable cone shrinks as distance increases.

When it came to the limits of both resolution and sensitivity, instrument designers quickly realized that bigger was better. Early telescope manufacturers lengthened their tubes to separate objective and eyepiece lenses, thereby maximizing magnification. When diffraction limits began to be appreciated, the diameter of telescope tubes also grew to accommodate larger lenses or, increasingly, mirrors (as the benefits of reflective telescopes were realized). The Gran Telescopio Canarias (GTC) on the island of La Palma in the Canary Islands boasts a mirror 10.4 meters (34 feet) in diameter. The twin Keck telescopes atop the Mauna Kea volcano in Hawaii are close behind at 10 meters each. In 2015, construction began on the 25.4 meter Giant Magellan Telescope, which is scheduled to begin observing from a high-altitude site in Chile's Atacama Desert in the early 2030s. Still more extreme are the massive parabolic dishes many people now associate with modern astronomy. The Arecibo radio telescope in Puerto Rico, built into a natural sinkhole in 1963, is 305 meters (1,000 feet) in diameter. As of this writing, the current record holder for the largest single-dish radio telescope in the world is the Five-Hundred-Meter Aperture Spherical Telescope (FAST) in the province of Guizhou, China, which began operation in 2016 and which measures a whopping 500 meters across.[8]

In addition to pushing the limits of angular resolution, large telescope dishes have the benefit of collecting a lot of light, which is essential for sensitivity. Also essential for sensitivity, as with microscopes, are sensitive detectors to record light, standing in for our naked retinas. To maximize signal-to-noise ratio, of course, it is also necessary to minimize noise. Visible-light telescopes like GTC and Keck were deliberately placed on isolated islands and mountaintops to limit light pollution from terrestrial sources. Earth's atmosphere absorbs light, so high-altitude locations, where the atmosphere is thin, were sought out to maximize light collection at certain more heavily absorbed wavelengths. Adaptive optics were also developed to overcome atmospheric distortions. Once rocket science got up a head of steam, it became possible to isolate telescopes even more effectively from noise, and at the same time remove them from the interfering effects of atmosphere, by launching them into space. It remains difficult even for twenty-first-century rockets to accommodate large payloads, hence the remarkable feat of high-tech origami that allowed the James Webb Space Telescope to unfold its tightly packed primary mirror to an eventual diameter of 6.5 meters, making itself as large as possible in order to deliver its much-vaunted resolution and sensitivity.

In chapter 5, I sketched out for you some of the origins of astronomical imaging in various parts of the electromagnetic spectrum. Suffice it to say here that different practical constraints on the collection and manipulation of light apply at different wavelengths. Countless improvements in materials science and engineering have made their way into modern telescopes, from one end of the spectrum to the other.[9] Different wavelengths also naturally result in different susceptibility to diffraction, so different flavors of astronomy have different intrinsic resolution limits. How have those limits stood the test of time? Did astronomers keep pace with microscopists in getting around the diffraction limit altogether? Most super-resolution techniques in microscopy require some form of intimate access to the imaged object: probing it with needles, surrounding it with lasers, or infiltrating it with activatable dyes. No such measures are available to astronomers. In the challenging case of radio astronomy, though, astronomers still managed to find a way.

In 1933, a Bell Laboratories radio engineer named Karl Jansky was looking for sources of noise that might interfere with radio transmissions. He discovered radio waves coming from the Milky Way, and from then on it was understood that the universe was whispering secrets to us on radio channels. When it came to pinpointing the sources of those radio emissions, however, there was a challenge to overcome. The wavelengths of radio waves are much longer than those of visible light—thousands to millions of times longer—so radio telescopes would need correspondingly larger diameters than those of visible light telescopes to deliver the same resolution. As impressively grand as the Arecibo and FAST radio dishes may be, they would need to be much grander still—on the scale of hundreds to thousands of kilometers—to approach the diffraction-limited spatial resolution of state-of-the-art optical telescopes. Nevertheless, starting in 2019, the Event Horizon Telescope, which is tuned to radio frequencies, somehow managed to achieve an angular resolution better than 60 microarcseconds—about the angular size of an orange on the moon as seen from Earth. This raw resolution is much finer than what is on offer from the eagle-eyed James Webb telescope in its special space aerie.

This feat was accomplished using a technique called interferometry. Interferometry has a long history in physics as a tool for high-precision measurement. It was used at the turn of the nineteenth century to establish that light behaves as a wave, and it is what enables the detection of gravitational waves today. In general, interferometry involves combining waves that have traveled along different paths. In the context of radio astronomy, it is accomplished by combining radio signals detected in separated collecting dishes and measuring the correlation of those signals in time with exquisite accuracy. Each measured correlation corresponds to a distinct projection of the arrangement of radio-wave-emitting objects in the sky, and an image is reconstructed through appropriate transformations of a set of distinct projections. I hope that, by now, I do not even need to point out the obvious analogy. This is tomography, yet again. The principles outlined in chapter 4 continue to apply.

Very-long-baseline interferometry (VLBI), as it has come to be called, got its start in the 1960s, spurred on by emerging computer hardware and fast algorithms.[10] The limit on spatial resolution in VLBI is determined not by the size of individual collecting dishes but rather by the maximum separation between dishes, otherwise known as the baseline. This new constraint provided a major boost to the "bigger is better" trend in astronomy. Large arrays of radio dishes became the norm. Consider the Very Large Array in Socorro, New Mexico. As its name suggests, the VLA is . . . very large. Its twenty-seven radio antennas each measure 25 meters (82 feet) across. Stretched out in a "Y" formation across a flat expanse of desert, they work together to achieve the resolution of an effective single antenna 35 kilometers (22 miles) wide.[11] If you thought that was big, meet the Very Long Baseline Array (VLBA). It consists of ten individual observing stations situated across 8,612 kilometers (5,351 miles), stretching from the U.S. Virgin Islands to Hawaii. Then there is the Event Horizon Telescope (EHT), shown in the middle of figure 6.3. In 2019, the EHT gave humanity its first-ever direct view of a black hole.[12] This groundbreaking image was obtained by digitizing and comparing signals from the Atacama Large Millimeter/submillimeter Array (ALMA) in Chile, an observing station at the South Pole, a station in Greenland, and a handful of other sites strategically

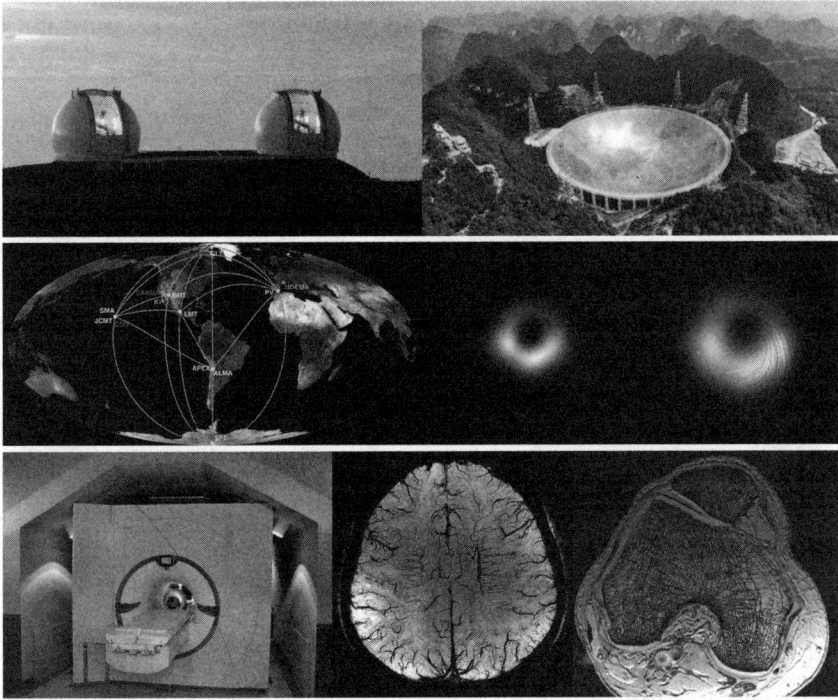

**FIGURE 6.3** Bigger is better. (*Top*) The twin Keck telescopes atop a volcanic peak on the Big Island of Hawaii (*left*) and the FAST telescope nestled among mountains in southwest China (*right*). (*Middle*) Observatories around the world contributing to the Event Horizon Telescope in 2017 and 2018 (*left*) and some of humanity's first direct images of a black hole (*right*). (*Bottom*) New York University's 7 Tesla MRI machine (*left*) and some early images (small blood vessels in a healthy brain; fine bony structures in the knee of a patient with osteoporosis).

*Sources*: (*Top left, top right*) SiOwl, CC BY 3.0 and Rodrigo con la G, CC BY-SA 4.0, via Wikimedia Commons; (*middle*) EHT Collaboration, CC BY 3.0 and 4.0, via Wikimedia Commons; (*bottom left*) courtesy of Joseph Helpern; (*bottom center*) courtesy of Yulin Ge; (*bottom right*) courtesy of Gregory Chang.

positioned around the globe. With its dispersed component sites operating in sync, the EHT was effectively the size of Earth itself.

Statements like this about effective size should be treated with some caution. Resolution limits in VLBI are often framed as "effective" diffraction limits, in which the diameter of a single telescope aperture is simply replaced by the maximum separation of observing stations. In actuality, though, diffraction is not the limiting factor in VLBI resolution. Like x-ray crystallography, VLBI uses wave behavior as an asset rather than a liability. VLBI's limits are therefore more tomographic than optical. The quality of a VLBI image depends on the number and the nature of available projections, which is given by the number and distribution of observing stations. The separation of observatory pairs is limited only by practicality. Nothing prevents us from combining Earth-bound detectors with detectors out in space, as long as the various signals can be collected and compared with sufficient accuracy.[13] Given enough time, effort, and sheer force of multi-generational planning, one could in principle create a telescope array with a baseline spanning the solar system or even the galaxy.[14] As for the number of observing stations, this is a matter of engineering, cost, and time. One trick long used by radio astronomers to make the most of limited numbers of observatories is simply to let time pass between observations. As the earth rotates, projections generated by detector pairs fixed on the planet's surface also rotate, filling in gaps in the measured projection data. In other words, rather than gathering images all at once like their optical cousins, radio telescope arrays typically scan in their image data over time. In this, and in a surprising number of other respects, radio astronomy bears a striking resemblance to MRI.

On the face of it, an MRI machine could not be more unlike an array of radio telescopes. One is a tube that fits in a room; the other is a regiment of giant antennas each the size of a house. One surrounds its living subject and makes intimate yet surgically precise slices without cutting; the other generates two-dimensional maps of a cold void populated by mind-bendingly distant objects. One looks inward; the other looks outward. The users of these two technologies also appear on the surface to have little in common. MR imagers explore millimeter-sized structures that change on timescales of milliseconds, whereas radio astronomers probe light-years of space over millennia of time. Radio astronomers and medical imagers generally attend different conferences, receive funding from different sources, and use different scientific jargon.

Despite these dramatic differences in scale and area of application, there are numerous deep connections. Like radio astronomy, MRI uses radio waves as an imaging probe. It therefore suffers from the same conundrum of spatial resolution. There were those who, before the 1970s, were absolutely certain that high-resolution imaging with nuclear magnetic resonance would be impossible because of diffraction limits, since the wavelength of radio waves typically employed in magnetic resonance experiments can be on the order of meters—just as big as the bodies one might wish to image. The skeptics were ultimately foiled by the power of tomographic transformations. Just as VLBI used interferometric measurements to generate projections of the sky, MRI used shaped magnetic fields to generate projections of the internal composition of bodies, and in both cases the resulting images were not subject to diffraction limits. In tomography, resolution comes down to how many projections can be gathered, and how distinct they can be made from one another. Not all projections can necessarily be gathered at once, so the process of data collection in tomography often takes time (which can also affect resolution when the imaged objects move). We will return to questions of imaging speed in chapter 8 and beyond.

Interestingly, just as for radio astronomy, the evolution of MRI technology since the 1970s has followed its own bigger-is-better trajectory. Admittedly, MRI machines don't communicate with one another as radio astronomy observatories do, and they don't yet bestride the world like a colossus. Still, some similar considerations of basic physics apply. The strength of magnetic field gradients in MRI is an analogue of the length of the baseline in VLBI, and field gradient hardware has advanced steadily, adding bulk and complexity to modern MR scanners. One key driver of the Human Connectome Project, which aims to map out the complex network of connections in the human brain, has been limit-pushing gradient technology. In recent years, colleagues of mine, led by Susie Huang, Larry Wald, and Bruce Rosen (one of the pioneers of functional MRI) at the Massachusetts General Hospital, have teamed up with Siemens Healthineers to develop Connectome 2.0 gradients, which are designed not just to map out neural circuits but to probe their cellular substrates at ever finer scales.[15]

The analogue for collector dish size, meanwhile, is the strength of the MRI magnet. If you want to see anything useful at all when you divide your tomographic signal up into smaller and smaller image pixels, you need higher and higher signal-to-noise ratio, and SNR in magnetic resonance is known to scale up with magnetic field strength. As a result, MRI has seen a steady march upward in magnetic field strength over time. The earliest imaging magnets operated well below 1 Tesla (a Tesla being a standard unit of magnetic field strength, named after the famous inventor Nikola Tesla). By the 1990s, 1.5 Tesla machines had become the standard in hospitals. Then 3 Tesla systems arrived, gobbling up significant market share. Since 2004, we have had a 7 Tesla MRI machine at New York University (figure 6.3, bottom). This machine—no 500-meter dish, but still something of a behemoth when you're standing in front of it—contains many miles of tightly wound superconducting wire encased in heavy-duty coolant tanks to keep the wire a few degrees above absolute zero. The whole thing is surrounded by 420 tons of steel that shield the space outside the magnet room from residual magnetic fields.[16] The bottom of the shield is embedded in concrete and anchored to bedrock to support the weight. The point of this whole rigmarole is to generate a magnetic field approximately 150,000 times as strong as the field that orients compasses on the earth's surface. When I'm trying to impress visitors unfamiliar with MRI, I sometimes call the 7 Tesla scanner our "refrigerator magnet," since it could hold up a refrigerator rather than the other way around. While 7 Tesla scanners are not in broad clinical use at present, some varieties have been approved for diagnostic purposes. Many sites around the world use 7 Tesla machines for research, and the hunger for sensitivity, spatial discrimination, and image contrast continues to drive field strengths still higher. Kamil Ugurbil, another fMRI pioneer and a long-standing champion of high-field magnets, operates a 10.5 Tesla machine at the University of Minnesota's Center for Magnetic Resonance Research. An 11.7 Tesla scanner at a facility outside Paris has recently released its first human images under the eye of the diffusion MRI innovator Denis Le Bihan. The National High Magnetic Field Laboratory in Florida has a 21.1 Tesla MRI machine that is too narrow for humans but can generate ultra-high-resolution pictures of animals. With dedicated detectors wrapped carefully around small samples, you can even use a massive machine like this as a microscope, resolving micrometers rather than millimeters.[17]

Size matters, then, in both radio astronomy and MRI, but the connections go deeper still. In addition to common physical constraints and analogous hardware elements, there are some remarkable similarities in the software used for imaging. These more striking similarities were brought home to me through a personal connection.

I first met Professor Urvashi Rau Venkata while preparing for a scientific meeting with the ambitious purpose of tracking the impact of imaging "from cells to galaxies." Urvashi is a radio astronomer. For the "Cells to Galaxies" conference, she was scheduled to deliver a lecture introducing medical imaging professionals to radio astronomy. My mandate was the opposite: to introduce astronomers to medical imaging. As educated scientists, we had perhaps some inkling going in that similar mathematical transformations were in vogue in our two areas of specialty, VLBI and MRI. When we first met by Zoom to coordinate our lectures and started going through our slide presentations, however, our jaws dropped. The math we had used to describe our respective processes of image reconstruction was just about identical. Apart from some minor changes in notation, we could hardly tell whose slides were whose. I would point to a term in one of my equations, and Urvashi would identify the corresponding term in hers. She would describe a challenge in calibrating radio telescopes, and I would exclaim, "We have that problem with our MR scanners, too!" Despite having absolutely no training in each other's professional discipline, we spoke a common language and had a common base of experience to draw from. It was like discovering a sibling you never knew you had.

Like the life story of a lost sibling, Urvashi's origin story as an imager also made a deep kind of sense to me. In much the same way that I came to MRI, she was drawn to radio astronomy by the transformations that make it work. She had always had an eye for the stars, and she found herself one summer in an undergraduate astrophysics program at India's Giant Metrewave Radio Telescope facility. After a lecture on the principles of radio astronomy, she and her classmates visited the telescope, where a revelation struck her: "This camera doesn't take pictures," she thought, "it *makes* them." The images displayed at the facility, she realized, weren't what-you-see-is-what-you-get propositions. They were the outcome of complex mathematical operations. The thought fascinated her.

After obtaining a master's degree in computer science—working, believe it or not, on simulations of cells—Urvashi found her way back to galaxies. She did an internship at the National Radio Astronomy Observatory (NRAO) in New Mexico, stayed to complete her PhD devising new ways of reconstructing images from radio telescopes, and was then recruited to join the NRAO staff. She has never left.

As part of the story of bigger-is-better astronomy, I've already made mention of the three major radio telescope facilities operated by the NRAO: the VLA, the VLBA, and ALMA.[18] Getting any actual images from these expensive facilities relies on software developed by Urvashi and her team.

Like the starburst of raw MRI data that you saw in chapter 5, raw radio astronomy data is generally uninterpretable by the eye. The data must be transformed to yield familiar pictures of celestial structures like the Crab Nebula. Urvashi leads scientific development for the Common Astronomy Software Applications (CASA) software platform. In addition to NRAO scientists, the CASA team includes scientists from the European Southern Observatory and the National Astronomical Observatory of Japan.

Urvashi's generally calm and composed demeanor belies the delight she clearly takes in building and sharing powerful imaging algorithms. At our first meeting, as we spoke about how we each went about separating signal from noise in our respective imaging modalities, her face lit up, as if her oval wire-rimmed glasses were focusing a source of internal illumination directly on me. You may think of her as a computational scientist if you like, but in my eyes she is a visual alchemist, discovering and skillfully executing the transformations that convert raw, leaden data into imaging gold.

Since the first "Cells to Galaxies" conference, I have had various opportunities to interact with Urvashi and with radio astronomy. I delivered another paired lecture at a follow-up conference with Urvashi's husband, Sanjay Bhatnagar, who is also an astrophysicist at the NRAO. At my invitation, Urvashi spoke at a New York conference devoted to the future of imaging, and she has become a go-to explicator of astronomy on the medical imaging conference circuit. I visited Urvashi and Sanjay at the NRAO, comparing notes and trading stories with them and their colleagues. While there, I toured the VLA, feeling small as I clambered partway up a towering dish in a repair hangar, and feeling part of something unaccountably large as I contemplated the ranks of active dishes disappearing into the sere desert distance.

As the scope of artificial imaging has expanded, its separate subcommunities have often lost touch with one another. This is clearly a missed opportunity. While some insights have certainly migrated from subfield to subfield over the years, how much more might we discover if we actually made a point of sitting down together regularly? What tricks from radio astronomy might shed light on how to make better MR images, and vice versa? What new and unexpected family relationships might we uncover? Case in point: the mathematics of structure determination in x-ray crystallography turn out to be closely related to how images are generated in both MRI and radio astronomy. Crystallography is yet another lost sibling, speaking the same language in perhaps just a slightly different dialect.[19]

If there is one thing I hope to have conveyed to you in this chapter, apart from the stubborn refusal of innovative imagers to accept their limitations, it is the remarkable and continuing connectedness of imaging. The fields of microscopy, astronomy, and medical imaging are all highly specialized, highly technical fields with very different nominal concerns. When looked at from another perspective, however, these fields are tightly linked. They confront similar limits and overcome them by dint of ingenuity or sheer gumption. They reconstruct their disparate images in sometimes unnervingly similar ways. Once they have images in hand, they also process them similarly, massaging their pixels to convey diverse information, as you saw in chapter 5.[20] Even as the tools of artificial imaging advance and multiply, the community of imagers remains connected in ways its members often do not appreciate. They share in an age-old human quest to augment our vision, yes, but they are also linked by common bonds of physics, engineering, and mathematics. Wildly dissimilar though the targets of vision may be, the instruments of vision lie closer together than one might think in the toolkit of human achievement.

# 7

# A Community of Imagers

A look at some of the people swept up in the modern medical imaging enterprise.

**M**y personal experience of imaging may be a little different from yours. For me, imaging is as much about community as it is about technology. My everyday world is populated by imagers. Even if you have a solid grasp of the principles by now, you may still feel somewhat on the outside if you are not a picture-making professional yourself. Allow me, then, to make a few introductions.

Imaging today encompasses many communities, and a comprehensive survey of any of them would be as difficult for me to distill as it would be exhausting for you to digest. Therefore, in this chapter I focus on a few of the people I have met in my travels. Since I have traveled most extensively in the areas of medical imaging and MRI, these are the areas that will be represented in the stories to follow. Please do not take this little sampling as any kind of attempt at a complete picture of the kaleidoscopic diversity of modern imaging.

Speaking of diversity, back in chapter 4 I promised you a counterpoint to the familiar inventor-hero model of scientific history in which progress appears to proceed from Nobel laureate to Nobel laureate. By now you have ample evidence of the contentious story behind many of the prizes we like to use as landmarks. Here is the backstory to that backstory: scientific progress is far more often a relay than an individual race. Scientific communities may be driven in part by competition, but if you step back from the buzz of individual rivalries, you can recognize these communities as geography- and generation-spanning teams, building idea upon idea and connecting one unsung hero to the next.

In chapter 6, in addition to providing you with a cavalcade of Nobel prizes in microscopy, I shared a serendipitous personal encounter with a radio astronomer. Here, I provide a curated sampling of character sketches you can use to take stock of the increasingly far-reaching human impact of medical imaging. Figure 7.1 provides a montage of faces corresponding to these sketches. For those of you who may be impatient to get on with the story of how imaging is being disrupted, feel free to skip lightly here, or even to jump all the way to chapter 8. For those who are interested in putting a human face to imaging, read on.

Roberta Kravitz is not just the former executive director of the International Society for Magnetic Resonance in Medicine (ISMRM). She is also a client. For years, nearly every moment of her professional life was occupied with imaging. A few years ago, imaging also saved her life.

In 2019, Roberta thought she had pulled something in her shoulder. Life as the executive director of a nonprofit is not exactly low-key, so Roberta ignored her complaining shoulder for a while and carried on with juggling the many demands of her job. The pain was annoying enough, though, that she eventually saw her doctor, who referred her for an MRI to help figure out what might be going on. When the images came through for interpretation, the radiologist reading the scan spotted an anomalous signal in her bone marrow. Roberta soon found herself queued up for a series of other imaging tests. It was a PET-CT scan that eventually clinched the diagnosis. Roberta had stage IV lung cancer that had metastasized to her shoulder.

She was rushed into radiation treatment. Thankfully, her treatment had the intended effect: the tumor shrank until it was no longer detectable on imaging. As of this writing, Roberta remains tumor-free. She is now a frequent-flier for follow-up PET-CT scans, which probe for any sign of anomalous metabolism that could hint at a return of the cancer. While Roberta does not relish those periodic scans, she sees them as a lifeline. She also credits that first MRI scan for giving her the critical early warning without which she would likely not be alive today. True, the rhythmic pounding of the MRI pulse sequences disturbed her attempts at meditation

**FIGURE 7.1** Dramatis personae: a worldwide community of imagers.

*Sources*: Courtesy of Roberta Kravitz, Mary Bruno, Robert Grossman, Roderic Pettigrew, Jürgen Hennig, Walter Märzendorfer, Michael Recht, Karla Miller, Urvashi Rao, Hedvig Hricak, Esther Warnert, Udunna Anazodo, Shy Shoham, and Aaron Sodickson. Photo of Graham Wiggins taken by the author. Blurred montage of unnamed imagers at bottom created by the author using DALL-E 3 (OpenAI).

in the scanner, but Roberta is not inclined to criticize. Had her tumor not been caught before it had a chance to metastasize more broadly, the aggressive therapy she received would have had a much lower likelihood of success. Roberta says there is something "round" about the fact that the modality that flagged her tumor was MRI—the imaging technique she has devoted much of her life to supporting.

Roberta, you see, is no ordinary consumer of medical imaging. She describes herself as an "MRI lifer." She found her way to medical imaging serendipitously through an early job in administrative support for the ISMRM, which happened to be headquartered near where she grew up in Northern California. She rose through the ranks, learning how to wrangle imaging researchers, how to organize large scientific meetings, and, eventually, how to oversee an efficient central office staff catering to the various and sundry needs of people engaged with MRI around the world.

The ISMRM was founded in 1994 as the result of a merger between two societies that had sprung up around the emerging field of MRI in the early 1980s. The society publishes two journals: *Magnetic Resonance in Medicine* and the *Journal of Magnetic Resonance Imaging*. It organizes numerous educational and networking opportunities for its membership of nearly ten thousand clinicians and scientists worldwide. It weighs in from time to time on best practices and unsolved problems in MRI. In a nutshell, its activities are dedicated to "extending vision, expanding minds, and improving life through magnetic resonance."[1]

Of all these activities, The ISMRM's signature offering is its annual scientific meeting. Roberta oversaw the logistical marvel that is the ISMRM annual meeting for many years, first as director of meetings and then as executive director. Over the course of a jam-packed week sometime between April and June each year, the meeting brings together many thousands of attendees for well over a thousand scientific and educational lectures, not to mention countless animated conversations in meeting rooms, in hotel lobbies, and on city streets. All this unfolds in and around capacious exhibition halls located in a wide network of countries that are selected for global reach (like Olympic venues but without quite as much politics). If you were to catch sight of Roberta at an annual meeting, she would likely be in motion, walkie-talkie in hand, fielding simultaneous questions about room setup and attendee registration while on her way to a boardroom for her seventh committee meeting of the day. Run into her at the closing party, however, and she would greet you by name (Roberta knows a remarkable number of names), hand you a drink ticket, and reminisce with a twinkle in her eye about old times, quirky colleagues, and dear friends.

The milestones of my own development as an imaging scientist are marked by the ISMRM annual meeting, from my first nerve-racking scientific presentation on MRI in a Vancouver convention center in 1997 to my stint as chair of the annual meeting program committee in 2010. The 2010 meeting took place in Stockholm, under the literal shadow of an Icelandic volcano whose eruptions had darkened skies and rerouted air traffic for several weeks. Roberta and I were in constant communication with the ISMRM board of trustees, debating whether it was safe to proceed with the meeting. (These were quaint, pre-COVID days, when stranded travelers were our principal concern.) We ultimately decided to go ahead, and the meeting opened with a welcome from the Queen of Sweden, followed by a reception at Stockholm City Hall, which hosts the annual Nobel Prize banquet. At one point during the week, as I recall, I was standing with Roberta on a high floor of the convention center, looking down on the seemingly unending streams of meeting attendees coursing through the center's network of hallways and open spaces. All these people, I remember reflecting, were here to advance health and understanding using imaging. This entire crush of collegial humanity was made up of acolytes of a single imaging modality that had appeared unexpectedly nearly four decades before. At the time, neither Roberta nor I gave much thought to the fact that among this surging crowd, there were surely experts on spotting anomalous signal in bone marrow that could be an early sign of metastatic lung cancer. Little did we know then that, one day, she would owe not only her livelihood but also her life to this community's collective inventiveness.

During the first few days of the ISMRM meeting each year, an affiliated society—the International Society for MR Radiographers & Technologists (ISMRT)—holds its own annual meeting in the same venue. Like the clinicians and imaging scientists with whom they work, members of the ISMRT also have an important stake in the development and use of medical imaging technologies. After all, they are generally the ones who actually make the images.

For some time after the discovery of x-rays, physicians operated their own x-ray equipment. But as the impact of x-ray images on health care grew and as the volume of imaging examinations increased, it became clear that dedicated specialists in image acquisition were needed. Thus was born the profession of radiologic technologist, or radiographer. The first professional association of radiologic technologists was formed in 1920.[2] As imaging technology advanced, imaging technologists advanced with it. The 1970s came and went, and radiographers specialized further to become experts in the operation of high-tech imaging tubes. As imaging studies grew ever more complex, and as radiologists retired to their darkened reading rooms to interpret a deluge of cross-sectional images, it was radiographers rather than radiologists who became the human face of medical imaging. Radiographers are the people who usher you into the scan room, bundle you onto the table, administer contrast material through an intravenous line if necessary, and talk you through your scan. They tell you when you need to hold your breath and when you can try to relax. They ask you what music you'd like to listen to during longer imaging sessions that are typical of, say, MRI. (My go-to artist when I'm in an MRI scanner is Ed Sheeran, whose syncopated rhythms mesh nicely with the banging and buzzing of the machine.) At the same time, these reassuring human presences must also be experts both in human anatomy and in device physics, since it is up to them to make sure the images come out right. Radiographers performed the scans that saved Roberta's life—radiographers like my colleague Mary Bruno.

Mary is a radiologic technologist specializing in MRI. She works in my home department, the Department of Radiology at the New York University Grossman School of Medicine. Hers is just the sort of voice you want to hear when you are lying alone in a noisy tube, wondering how you got there, and trying to imagine what comes next. Mary is kind and engaging, reassuringly calm but also unflaggingly upbeat. She also knows MRI machines inside and out. She is techno-magician, psychotherapist, and fellow-feeler all rolled up into one.

Mary is a second-generation imaging professional. Her father was working in security at a hospital when he fell into conversation with an x-ray technologist and promptly caught the x-ray bug himself. When Mary was about seven years old, her father switched to MRI, which was then just beginning to make its way into clinical practices. Mary grew up on a steady diet of images that her father would bring home to show the family. He was captivated by what these images showed, and so was she.

Mary is a people person who loves science and feels at home in fast-paced environments. In college, she explored forensic science but found that she preferred direct interpersonal interactions to lab work, so she enrolled in an x-ray program. When NYU recruiters paid a visit looking for an MR technologist, she jumped. In her first year at NYU, a neuroradiologist she worked with showed her early fMRI images that were being used to plan brain surgeries, and that sealed the deal for her. She saw that there was exciting new science to do, behind the medical images that had accompanied her through life.

Today, Mary splits her time between clinical work, research, and education. She continues to be a people person, she continues to love science, and the environment in which she works continues to be unrelentingly fast-paced. On a typical clinical day, she may scan fifteen patients in a row. Let's say you are one of them. She will greet you with a smile, take time to put you at ease, and then set you up, safe and comfortable, in the MRI tube. As soon as you are in the scanner, Mary becomes a one-woman *Star Trek* Enterprise bridge crew, fingers flying across keyboards, juggling display screens packed with arcane parameters and annotated images, all while maintaining periodic encouraging contact with you. She adjusts imaging protocols on the fly, using her encyclopedic knowledge of MRI physics to boost imaging speed if you are restless and can't hold still, or choosing a specialized sequence to image safely around the implant in your hip. Then, when all the mechanical humming and buzzing is done, she pulls you out of the tube, answers your questions, makes sure you understand what you need to understand, reunites you with the non-magnet-safe belongings you removed before the scan, and sends you on your way with a smile.

Your time in the imaging suite may be done, but Mary's is just getting started. As an advanced practice specialist charged with education and mentorship, she may be monitoring the work of two junior technologists on two other scanners, offering advice or joining in to help as needed. She may be battling the clock to minimize delays in scanner schedules that run from 6:45 A.M. to 11 P.M., seven days a week. She may be covering for the multilingual technologist who is triple-checking the translation of a report to make sure that it will make sense to a patient whose first language is not English. On an academic day, or after hours, Mary will study up on recent literature to prepare monthly educational lectures for her staff—say, on how best to image metallic implants. Or she may take some time to familiarize herself with the latest update of the ever-changing operating software running NYU's MR scanners. Then there is research. You will often find Mary sitting at the scanners with imaging researchers like me, juggling the constraints of physics, physiology, and human behavior to come up with research studies that will actually work in practice. She recruits human subjects for such studies and performs specialized research scans with brand-new detectors and experimental software. She also serves on the abstract committee for the ISMRT, evaluating original research by MR technologists around the world. Mary is a part of many worlds and many teams. If ever you find yourself in need of medical imaging, you will want her on your team as well.

❧

The NYU Grossman School of Medicine where Mary and I work is named after its long-serving dean, Dr. Robert Grossman. Bob is a radiologist. His sharp eye for anomalies in hospital operations was honed by hunting for anomalies in images. Bob has always been a visual kind of person. He is quick to see patterns where others might not. Were it not for a quirk of history, though, his career in imaging might never have materialized.

Bob is the mentor I mentioned back in chapter 4 who thought he was going to become a brain surgeon until medical images intervened. As he tells the story, he was in the last year of his neurosurgery residency in 1973 when he was given the opportunity to attend a conference on a new-fangled technology called computed tomography. The images Bob saw at the conference were of remarkably low resolution by today's standards, but they were all it took to move him. He came home convinced that the next advances in neurosurgery were going to come through imaging. He told his wife, "I'm going to change careers and go into radiology." Bob's father was nonplussed on hearing the news: "Let me get this straight. You're giving up brain surgery to take pictures?"

What was so convincing about those early CT images? How could Bob be so certain that imaging was the future of brain science and medicine? He could be so certain because, for the first time in his experience working with the brain, he felt that he could see precisely what he needed to see before he started cutting. He had been trained to puzzle over skull x-rays, which was a little like reading tea leaves. He had watched surgeons try to work out the location of abnormalities on angiograms (x-ray images enhanced with high-density dye injected into patients' veins to highlight blood vessels), and it reminded him of scholars bobbing over inscrutable texts. Then there were the more invasive imaging procedures like myelog-raphy (in which dye was injected directly into the spinal canal) and pneu-moencephalography (in which cerebrospinal fluid was drained from the brain via a lumbar puncture and was replaced with air, which was more visible on x-ray images). To Bob, these seemed like barbaric ways to go about enhancing image contrast. Contrast could be found in abundance in the CT images, however. Hemorrhage—bleeding in the brain, which could be a life-or-death finding for a patient—showed up clear as day. Previously, Bob would have had to deduce the presence of blood through subtle shifts in faint brain structures. In the CT images, it was just right there in front of his eyes.

As a radiologist, Bob ended up developing a serious taste for blood. He was finishing his radiology training when the first MR images came out, and, sure enough, they were also sensitive to the presence of blood in the brain.[3] Bob devoted himself to working out the changing appearance of blood at various times after a brain injury is incurred. This informa-tion proved to be quite valuable both for diagnosis and for the guidance of therapy, and blood-sensitive pulse sequences remain a standard feature of many modern brain MRI protocols. Working with progressively more advanced MRI machines, he also cataloged subtle imaging characteristics of multiple sclerosis and other neurological disorders. In 1994, he pub-lished a textbook on neuroradiology.

Bob came to NYU in 2001 as chair of the Department of Radiology. It was he and his vice chair for research, Dr. Vivian Lee, who recruited me to NYU in 2006. Not long afterward, Bob was appointed dean of the NYU School of Medicine and CEO of its medical center. He has presided over a marked expansion of the medical center's reach and reputation since then. To those who know Bob, this came as no surprise. He has sky-high stan-dards and a keen instinct for competition. He is not somebody you want to disappoint. Don't even think about trying to pull the wool over his eyes, either, if your news isn't good. Remember—he sees patterns in the data even if you might not want him to. Bob is not afraid to ruffle feathers. At the same time, it may surprise you how generous he is with his time, and how devoted a friend he can be. He loves to hear about what's new, in sci-ence and in life. Get him talking about the old days in imaging, on the other hand, and he'll open up like a book, leaning back in his chair and holding forth with what can only be described as glee.

It is not difficult to draw a line between Bob's exacting standards for image quality and his uncompromising emphasis on data-driven quality of care as a leader in health care. His whole career was built on the idea that if you can see it, you can cure it.

"There are two things in life everyone wants more of. One of them is SNR."

The author of this memorable, if slightly inscrutable, quote is another noted leader in the overlapping spheres of imaging and health care. He is also someone I am proud to claim as a mentor in my own imaging journey. His name is Dr. Roderic Pettigrew.

Rod was born in Georgia in 1951. He gobbled up stories of the latest scientific advances as a child, gravitated toward science in his studies, and went on to amass a stellar scientific résumé: a BS in physics from More-house College, an MS in nuclear science and engineering from Rensselaer Polytechnic Institute, a PhD in radiation physics from the Massachusetts Institute of Technology, and an MD from the University of Miami in an accelerated two-year program specifically tailored to physician scientists. As he continued his medical training at Emory University and then at the University of California San Diego (UCSD), he found a special place in his heart for imaging. In his nuclear medicine residency at UCSD, he specialized in SPECT (a cousin to PET imaging that had shown particular value for cardiac examinations). He joined Picker International as a clinical research scientist in 1983, just as the company was setting out to develop MRI specifically for the heart. Rod wrote the software that allowed Picker's first MR scanners to image the cardiovascular system. He personally installed and calibrated this software on the first ten Picker systems worldwide and instructed clinicians on its use. Rod then returned to Emory to take up a faculty position. At Emory, he built a reputation as a leading expert on cardiovascular imaging. When Philips Medical Systems set out to develop their first industrial cardiovascular MRI products, Rod was their partner, guiding the development of the first dedicated cardiac MR image display and analysis software package. Rod was selected to deliver the New Horizons Lecture of the venerable Radiological Society of North America (RSNA) in 1989—the RSNA's seventy-fifth anniversary year—and he spoke about 4-D cardiac MRI as the diagnostic procedure of the future.

He headed the Emory Center for Magnetic Resonance Research and garnered professorships in both cardiology and bioengineering.

Throughout this meteoric rise to the upper echelons of the imaging elite, Rod had a consistent challenge to contend with: he didn't look like other imagers. Rod is Black. As an imaging scientist, Rod is intimately familiar with the origins and manipulations of color. He knows full well that color is literally a matter of perception. He has also learned through long experience that people can be blinded by preconceptions when they look at him.[4] One day in the late 1950s, while the American civil rights movement was still gathering steam, a youthful Rod—only five or six years old at the time—was innocently (and rather precociously) perusing an issue of *Popular Science* in a grocery store. An angry onlooker, offended by what he saw, confronted Rod with racial slurs. Rod was young enough to have no idea what the N-word meant, but its intent was all too clear. He still recalls his shock at the anger, the hatred, the unaccountable meanness in the face of this stranger. Years later, when asking for directions to the MRI suite in a hospital, Rod was told by a well-known cardiologist that he needn't be asking for the MRI room since the room didn't need cleaning, thank you kindly. As it happens, Rod was there to install his new cardiac imaging software on the MRI system. He would later teach that same cardiologist— the one who had immediately assumed that a Black man would be there to sweep up after him—how to use the advanced software Rod had built. Rod never let aggressions like this stop him. Their collective drag is something I never had to push against as I made my way into imaging, though.

It was in the late 1990s at a workshop on cardiac MRI—by then a flourishing field of study—that I first met Rod and was treated to his opinions about the universal desirability of signal-to-noise ratio. On display at that same workshop were abundant eye-catching snapshots of the heart in its natural habitat inside the chest. Lacking the shutter speed to freeze out cardiac motion altogether, ingenious researchers had figured out how to piece together short exposures over multiple repeating heartbeats to build up a high-resolution composite picture. Other researchers had assembled sequential snapshots into detailed movies of the beating heart. In these movies, one could track the subtle twists and turns of contracting hearts and follow the blood they ejected as it coursed throughout the body. It was mesmerizing.

In 2002, Rod was recruited to be the founding director of the National Institute of Biomedical Imaging and Bioengineering (NIBIB). Did you know that one of the U.S. National Institutes of Health is devoted to imaging? The

NIBIB was founded to catalyze a convergence of engineering, life sciences, and physical sciences research, including both biomedical imaging and bioengineering, which by the turn of the twenty-first century had spread widely through a large number of disease-focused institutes, including the National Cancer Institute, the National Institute of Neurological Disorders and Stroke, and the National Heart, Lung, and Blood Institute, among others. Rod ran the NIBIB for fifteen years before stepping down in 2017 to become the CEO and executive dean of a new medical school built with engineering at its core. This new school—the Texas A&M School of Engineering Medicine, or EnMed for short—brings engineering and medicine together in the same four-year curriculum. All students graduate with two degrees: an MD and a Master of Engineering. In order to graduate, students are required to innovate a solution to a health care problem. Rod has a name for the innovation-minded physicians emerging from EnMed; he calls them "physicianeers."[5]

Connections between engineering and medicine are ubiquitous in Rod's now full-to-overflowing résumé. (Case in point: he is an elected member of the National Academy of Medicine, the National Academy of Engineering, the National Academy of Inventors, the American Academy of Arts and Sciences, and even India's National Academy of Sciences.) Though imaging is only one piece of those ever-expanding connections, I like to think of it as a foundational piece. What is imaging, after all, but the engineering of all that we can see?

During his tenure at the NIBIB, Rod served as an ongoing advocate for imaging and engineering within the National Institutes of Health (NIH). At the time, the NIH was well versed in supporting research on major diseases, but its mandate was not quite as clear when it came to developing key technologies or other innovations that might one day advance biomedical knowledge and clinical care. Rod testified to Congress and spoke to audiences around the world about advances in imaging and engineering. This aspect of his role definitely played to his strengths. Rod collects stories of engineering triumphs, and he doles them out with relish. Not one to rush a punch line, he spins his tales in a gentlemanly drawl that is nevertheless particular and precise. Whenever a slide of mine has made it into one of his talks, he has inevitably given it more time and attention than I would have done myself.

A generation of new-tool developers, myself included, has by now received support from the NIBIB, and a number of "medical moonshots"

(Rod's term) have ensued. In addition to new state-of-the-art imaging methods using a wide range of probes, NIBIB-funded research has brought us cutting-edge methods of spinal cord repair, key steps toward the creation of artificial organs, home COVID-19 tests, and more. Continuing Rod's moonshot analogy, you can think of the NIBIB as an inward-looking counterpart to NASA. While NASA administers the U.S. space program, the NIBIB runs America's *inner*-space program.

Of course, the interest in new tools to help us see and treat disease is by no means limited to the United States. The charge has also been taken up by funding organizations in many other countries—including, say, Germany, where Jürgen Hennig has built his illustrious career.

When Jürgen and I stand next to each other, people laugh.

I'm not kidding with you. We have actually done the experiment. It was 2016, and Jürgen was chairing the Gordon Research Conference on In Vivo Magnetic Resonance. I was his vice chair: his assistant in putting the meeting together and his successor for the following meeting. Jürgen called me up to the front of the rustic conference hall in Andover, New Hampshire, so that we could stand shoulder to shoulder to address a crowd of imaging colleagues. The problem is that I stand a little south of five feet, six inches tall, and the top of my head doesn't reach Jürgen's shoulder. At closer to seven feet tall than he is to six, Jürgen towers over most people. Being a man accustomed to making an entrance, he paused for effect—just for a moment—as we faced the audience together. The contrast in our statures struck some primal visual chord, and the room erupted in laughter. Jürgen draped an arm around my low-lying shoulder, and we joined in.

Setting aside his sly sense of humor, what Prof. Dr. Jürgen Hennig does for a living is serious business. It is no accident that he knows how to make a visual impression, since he has been doing just that professionally since the 1980s. (You met him in passing back in chapter 5, actually, when I alluded to his invention of the fast RARE technique that helped establish one of the key image contrasts for modern MRI protocols.)

Jürgen began his scientific career as a chemist. His studies took him to Stuttgart, London, Munich, Freiburg, and Zurich. He had been investigating chemical reactions using combinations of light and nuclear magnetic resonance when the advent of MR imaging set him on a new path. He ended up back in Freiburg, doing experiments on the fourth MRI machine to be installed in Germany. He first developed his RARE technique in 1984. He was in attendance at a 1985 meeting on MRI (organized by one of the professional societies that would later merge into the ISMRM) when Raymond Damadian took advantage of his position as session moderator to rail at length against Paul Lauterbur. Jürgen recalls that Damadian caused enough of a stir to have his membership in the society revoked after the meeting. Jürgen, on the other hand, was just getting started. At a bachelor party in the mid-1980s, he was approached by a connection who worked at the German company Bruker, which had recently gotten into the MRI business. Soon afterward, Jürgen found himself in China, installing a scanner that produced the first MR image ever to be obtained in that country.[6] He has maintained warm personal and professional connections with China ever since.

As the 1980s wound down, Jürgen settled in back at the University of Freiburg in Germany. There, hc progressed from researcher to professor to director, ultimately building one of the most respected MRI research groups in the world. He served as president of the ISMRM from 1999 to 2000. He has trained a long lineage of students, some of whom have gone on to lead their own respected imaging research groups. By the time the 2016 Gordon Conference came around, he had become more or less a living legend in the MRI community. It didn't hurt his standing in the slightest when, toward the end of the meeting, he treated us to a live performance on the alpine horn, a straight, tapered tube longer than he is tall, traditionally used as a signal instrument in village communities of the Swiss Alps. Ever the pragmatic problem-solver, he had packed his alpenhorn up in short segments for the flight from Germany.

Jürgen is known for the clinical impact of his early contributions, to be sure, but also for his team's ongoing contributions to the evolving hardware and software of MRI machines. A tool-builder par excellence, Jürgen has applied those tools for the sake of basic scientific discovery, working with neuroscientists to characterize the structure and function of the brain, and joining forces with physiologists to generate pristine images of animal anatomy. He has combined MRI with other imaging modalities to tease out subtleties of living biology. His body of work represents an informative microcosm of what imaging researchers can do.

One thing researchers like Jürgen are continually called upon to do is to procure funding. With a lab that grew to include dozens of students and staff, Jürgen had many minds and mouths to feed. He became an expert in writing grant proposals to the German Research Foundation (Deutsche Forschungsgemeinschaft), the European Union, and other organizations willing to support imaging research. While many things separate the academic enterprise from industry, I have often reflected that a modern scientist must take on some of the characteristics of a CEO in order to survive in an increasingly competitive scientific landscape. If the idea of university-based research conjures up soft-focus images of pipe-wielding professors quietly luxuriating in their tenure, think again. Fundraising, recruitment, retention, public relations—all of these are essential skills for today's lab heads. Sometimes, all the associated strategizing and storytelling can help to hone important research directions. At least as often, these preoccupations stand in the way of scientific progress. In any case, the connection between academia and industry is more than academic when it comes to work like Jürgen's. When you are operating at the forefront of a high-technology field like MRI, direct interactions with imaging companies can also be critical.

❧

Four hundred kilometers northeast of Jürgen's Freiburg haunts lies the Bavarian city of Erlangen. Erlangen is a hub for Siemens, the electronics giant founded by the German inventor Werner von Siemens midway through the nineteenth century. Siemens got into the imaging game all the way back in 1896, manufacturing x-ray tubes for early forays into medical diagnostics in the first year after Röntgen's discovery. When the cross-sectional imaging craze hit seventy-five years later, Siemens jumped right in again. The first Siemens CT scanner launched in 1975. By 1978, a prototype Siemens MRI machine had produced an image of a bell pepper.[7] In 1980, Siemens engineers generated their first full-body human MR scan, and the first clinical MRI machine from Siemens was installed in 1983.

Walter Märzendorfer got into the Siemens game in 1984. After earning his diploma in electrical engineering, he had been working for just a year and a half as a faculty member of the Institute for Physiology and Biocybernetics at the Friedrich-Alexander University in Erlangen, when his chairman asked him if he would be interested in consulting with a design team at the neighboring Siemens plant. Walter had not yet heard about the emerging technology of MRI, but when he first cast eyes on the design team's prototype, it was love at first sight. (Nine hulking cabinets of electronics and rows of imposing transistor banks might not have been everyone's cup of tea, but technology has long been the way to Walter's heart.) He took a job at Siemens as a research engineer, responsible for handling the output of biosensors like EKGs that had to work in the presence of lots of other interfering systems, including an unforgivingly strong magnetic field.

Walter worked his way up the ranks at Siemens. In 1994, he was put in charge of research and development (R&D) for a new line of MR scanners that would come to be called "Magnetom Symphony" scanners, and that would be followed in short order by euphonious successors like "Sonata" and "Harmony." Feeling his oats as head of MR systems engineering, Walter opined to colleagues that, by comparison, CT was dead. As a reward, he was appointed head of R&D for the CT business unit in 2001, just in time to preside over a flowering of so-called multislice CT technology. In 2006, Walter got a call offering him what he still refers to as "the greatest job on Earth": CEO of the global MRI business unit at Siemens. The business units for individual imaging modalities at Siemens operated essentially as free-standing companies, and Walter had full responsibility for profit and loss, manufacturing, marketing, service, and of course R&D. In his five years as CEO, he oversaw the development of several workhorse MRI systems that are still in broad use around the world. He debated new directions in MR with thought leaders in the ISMRM community, set innovation strategy for thousands of employees, and optimized the operation of factory floors spanning many city blocks. In 2011, Walter was called back to CT as CEO of the CT and radiation oncology business unit, where he set strategy for both diagnostic and therapeutic uses of radiation. After four years in that role, he took over from his boss to become president of diagnostic imaging. Between 2015 and 2018, Walter oversaw the entire Siemens imaging portfolio.

Let us briefly take stock of that portfolio, so that you can get a sense of the current scope of commercial medical imaging. Certainly the four top-tier tomographic imaging modalities—CT, MRI, PET, and ultrasound—are well represented. Other nuclear and molecular imaging modalities are in the mix as well. Then there are imaging devices tailored for radiation therapy, surgery, and urology. And don't forget x-ray machines, which nowadays come in many varieties: simple plain-film devices that are descendants of the old x-ray machines, but also mammography devices, mobile machines for imaging in the operating room or at the bedside, and assorted other purpose-built modern instruments. Though many of the factories that produce this panoply of imaging devices are located in and around Erlangen, Siemens maintains other production plants around the world, as well an extended global network of service, marketing, and collaboration centers.

I told you that modern medical imaging is big business. Just how big is it, though? Together, the Siemens diagnostic imaging business units under Walter's purview were responsible for sales of thousands of imaging machines each year, as well as ongoing service for an install base measured in the hundreds of thousands. External revenue in the last year of Walter's tenure amounted to over €8 billion ($9.6 billion). By some estimates, the number of patient interactions with Siemens imaging technologies ran into the hundreds of thousands *per hour*.

Then there is the competition. Like Siemens, General Electric and Philips have been stalwarts in the medical imaging business from its early days. These three big horses in the imaging device race have historically traded off the lead in market share depending on who happened to have the latest and greatest in innovation or marketing. Various other companies, large and small, have made inroads over time. Some, like Hitachi and Toshiba, have maintained a steady second-tier presence with significant regional sales but more limited global reach. Others, like Cleveland-based Picker or the Israeli company Elscint, have been acquired by existing players. Still others, like Hewlett-Packard and the original CT manufacturer, EMI, have gotten out of the medical imaging business altogether. Assorted newcomers are currently vying for market share. This is just the sort of competitive landscape one might expect for a high-tech business driven by macrotrends in health care economics. In 2020, the annual global market size for medical imaging devices was estimated at nearly $40 billion.[8]

I got to know Walter starting around 2007, not long after he began his stint as MRI CEO. I had just joined NYU, whose research partnership with Siemens had been ramped up significantly during Bob Grossman's term as radiology chair. In addition to maintaining bustling in-house research programs, which birth new scanners and give facelifts to old ones on a regular basis, Siemens and its industry peers play a key enabling role in medical imaging research at large. Academic centers—even well-equipped ones like Jürgen's, or like NYU's—have traditionally been light on the manufacturing expertise required to build complex whole-body imaging systems from scratch. Industry, on the other hand, is generally removed from where the action is in basic biomedical science and clinical applications. So, the two sides have a history of teaming up to bring cutting-edge innovations into practice.

Such a team-up gave me a seat at the table with Walter for conversations about new directions in the science and practice of imaging. From the vantage point of a conversation partner seated across a table, Walter's features are a study in sharp lines: an aquiline noise framed by bold brows above rectangular half-rim glasses. His piercing eyes and ever-present hint of a smile give him an appearance of eager alertness, as if he is poised to pounce. His preferred quarry for pouncing is ideas. When he finds a good idea, those shrewd eyes of his are quick with a good-natured glint.

I distinctly remember the first tour Walter gave me of the Siemens MRI factory in Erlangen. I had never seen so many expensive machines all together in one place. There were naked tubes the length and width of a human torso suspended from tracks in the ceiling. There were thicker tubes hand-wound with wires in looping patterns like a giant's fingerprints etched in metal.[9] These tubes were trundled into great metal tanks housing giant spools of superconducting wire that would eventually be energized to create intense magnetic fields. Further along, motorized tables were docked to the growing structures, which soon blossomed with festooning cables and cabinets full of electronics. Then, last of all, came the skin of neat plastic trim emblazoned with Siemens logos. It was like a time-lapse movie of assembly spread out in successive snapshots across the factory floor. Or perhaps it resembled a stop-motion film of a developing child. Walter strode amongst the juvenile imaging machines like a proud papa, greeting the machines' human attendants with a familiar nod, and explaining all the stages of development with an affable grin.

The Siemens factory floor was also the subject of a different set of animated conversations in which I played the role of spectator from time to time. Walter's conversation partner on these occasions was my current radiology department chair, Dr. Michael Recht. Walter and Michael liked nothing better than discussing the similarities between operating a high-tech manufacturing facility and running a busy radiology department.

Michael knows the business of radiology inside and out. He came to NYU from the Cleveland Clinic in 2008 as Bob's chosen successor as chair of radiology, at which point he appointed me his vice chair for research. Like Bob, Michael has a keen eye for pathology, whether anatomical or operational. Unlike Bob, Michael specializes in musculoskeletal radiology. He knows knees and shoulders backward and forward. He carries with him a detailed mental map of all the comings and goings of joints and ligaments and tendons and muscles. When he first entered medical school, he was convinced that he would become a pediatric endocrinologist, but a radiology elective in his third year set him straight. He trained first in interventional radiology, doing therapeutic procedures under image guidance. Then MRI got a hold on him, and he changed course again. He actually spent ten months at Siemens' MR headquarters in Erlangen during a fellowship year, learning the technical ins and outs of MRI and providing clinical input to engineers there, before returning to the United States to specialize further in musculoskeletal imaging.

In addition to his knack for finding torn ligaments and subtle abnormalities in knee cartilage, Michael has an uncanny ability to spot a single suspicious entry in a budget spreadsheet full to the brimming with numbers. To be frank, I find this talent as infuriating as it is awe-inspiring. I'm a physicist, and by all accounts I'm pretty good with numbers. But Michael can reduce me to dumb puzzlement by jabbing a finger at one out-of-whack number that has caught his eye. It generally takes me several painstaking minutes of on-the-fly calculation to catch up.

As a leader, not to mention as a conversationalist, Michael is kinetic: always moving, always thinking, always planning his next move. He will tell you what that next move is—no agenda is hidden with him—but you will still find it difficult to keep up. We joke that his job is to speed me up, and my job is to slow him down. Michael and I have talked for countless hours about research directions, recruitment strategies, innovations in imaging education, and the future of radiology. We have worked together on using artificial intelligence to produce better and faster images of the knee. We have traveled together to Erlangen, to ISMRM annual meetings, and to the largest medical conference in the world: the annual Chicago-based conference of the RSNA.

That's right. For significant swaths of the twenty-first century so far, imaging has held the title for the largest medical meeting. In its prepandemic heyday, the RSNA annual meeting brought together more than fifty thousand people from more than 140 countries. In the words of one regular attendee, it's like "Disneyland for Radiologists."[10] As compared with the scientific sessions at the ISMRM annual meeting, RSNA sessions are focused more on the evolving clinical practice of imaging than on the development of new tools per se. RSNA educational offerings attract plenty of customers, and the plenary lectures pack in quite a crowd, but it is the exhibit halls that are the beating heart of the RSNA meeting, pulsing with the lifeblood of large-scale imaging. There, vendor booths are multi-story indoor cities, complete with eye-catching electronic billboards, life-sized mockups of imaging machines, and besuited sales reps hawking their wares. Think *Blade Runner* with better lighting. In popup meeting rooms custom-built into the backbone of this high-tech bazaar, CEOs like Walter and department heads like Michael conduct business, making deals, cementing relationships, and determining the future distribution of major medical imaging devices. In addition to the core device manufacturers, you will find a whole ecosystem of supporting industries on the RSNA exhibit floor: image archiving companies and electronic health record companies, vendors specializing in radiologic workflow management, and a slew of companies selling ancillary hardware and software to get the most out of imaging devices. You can take stock of technology trends year by year just by strolling through the aisles. You can also track evolving market forces.

This brings me back to the business of imaging. What is the value proposition of seeing inside ourselves? Clearly, it must involve some balance between costs and benefits. The benefits are obvious to anyone whose life has been saved by imaging and to any clinician whose patient's life has been saved. At a population level, however, the value of information gleaned from imaging may be more difficult to quantify. One can tally the number of years of life gained, the number of days of productive work restored, or perhaps the number of therapies appropriately modified based on image information. A whole discipline called comparative effectiveness is devoted to such tallies. And what about the costs? The price of an imaging machine typically ranges anywhere from $100,000 to multiple millions of dollars a pop. Then there are the ongoing fees for maintenance and service, which are nontrivial for large pieces of complex equipment. Radiologists, radiographers, and all the associated support staff needed to run imaging facilities have to be paid, of course. And someone needs to keep the lights on, the floors swept, and the machines fed with power.

In order to stay solvent, imaging departments must charge fees to offset all these costs. Various countries take various approaches to such fees. In some nations, like the United Kingdom, there is a high degree of centralization, with most fees being fixed by the National Health Service. In other countries, like the United States, a much larger proportion of imaging fees is paid by private insurance companies, and rates used by public insurers like Medicare serve only as benchmarks. Actual reimbursement rates for any given imaging exam are negotiated based on the leverage a particular medical system might have with a particular insurer. Debates have been raging for years about whether the future of health care should bear any resemblance to either the highly centralized UK model or the Wild West U.S. model. A number of experts have argued that we should abandon the current fee-for-service model altogether in favor of a value-based model, with levels of payment determined by patient outcomes.

However you do the cost–benefit analysis, one thing is clear. Until relatively recently, countries with access to advanced medical imaging have used it more and more. The dramatic rise in utilization of medical imaging has in turn contributed to a rise in health care costs. As a result, recent years have seen a growing pressure to decrease costs. The eyes of governments and insurers have increasingly turned to imaging as a cost-driver, and many have dropped their reimbursement levels for imaging examinations.[11] This has placed an ever-greater premium on efficiency for imaging enterprises.

Fortunately for the NYU radiology enterprise, "efficiency" is more or less Michael Recht's middle name. Sit with him in his office, and he will enthusiastically walk you through all the dashboards built to his specifications by his information technology team. At any given time, he can tell you exactly how well any imaging device at any affiliated facility is running. If a trend in the data does not look good, he will dispatch a team to fix it. If you can see it, you can cure it, right?

The result of all this operational engineering has been a marked expansion of imaging volume at NYU. When Michael started as chair in 2008, the department conducted about 350,000 imaging exams a year. By 2024, that number had risen to more than 2.5 million exams a year. Increased efficiency in existing imaging facilities has accounted for only part of this rise. The rest has resulted from a dramatic expansion in NYU's portfolio of imaging sites, which was achieved by taking on sites that had been struggling to make ends meet and remaking them in the new efficient model. At current count, Michael's dashboards cover more than fifty outpatient facilities and five hospitals. Each facility typically holds multiple imaging devices, so the device tally is substantially larger still. NYU's extended network includes all of fifty-five MRI machines, for example (with fourteen more scheduled to be installed in the next year or two). Supporting all these facilities are 275 radiologists and nearly 750 technologists. Including all the staff who manage billing, facilities planning, IT, and administration, plus a research team of 150, there are about 1,500 employees in the department. The annual departmental budget exceeds $450 million. Not all radiology departments have quite this scope, but I think you have seen enough by now to appreciate that, even at the level of individual medical centers, imaging is big business. When you add up all the facilities around the world that do medical imaging, imaging services are believed to account for the equivalent of hundreds of billions of U.S. dollars in annual expenditures.

I understand, of course, that when you yourself need to go in for an imaging exam, you are generally not thinking about all of this. That is why I'm giving you a few glimpses of the bigger picture now. As this chapter has progressed, I have tried to take you from the individual experience of imaging all the way up to the broad-brush perspective of organizations, macroeconomics, and populations. It is important to remember that imaging touches individual lives in many different ways. It is also important to remember that, taken together, the number of lives affected by imaging runs into the billions.

Estimates by the World Health Organization and other sources place the number of diagnostic imaging examinations performed globally each year at anywhere from 3.6 billion to more than 5 billion.[12] Given this huge volume of examinations, and given the rich information content of each exam, it stands to reason that imaging has much to teach us about the evolving health of the human population at large. Historically, though, medical imaging data has not been organized proactively on a large scale. Professor Karla Miller is doing her best to change that.

Karla is the associate director of the Wellcome Centre for Integrative Neuroimaging at the University of Oxford. She helped establish the brain imaging protocol for the UK Biobank, and she has led the way in making sense of the pictures that have emerged from it.

According to a recent online description, "UK Biobank is a large-scale biomedical database and research resource containing de-identified genetic, lifestyle and health information and biological samples from half a million UK participants."[13] Participants are followed over time with sequential collection of biological samples and ongoing biomedical tests, including imaging in a subset of one hundred thousand participants. The data gathered into the biobank is made available to researchers worldwide to foster new discoveries that may improve human health.

Why fuss over a hundred thousand or even several hundreds of thousands of participants in a single country when we know that medical data is obtained all the time for many more people than this? First of all, traditional medical data is not easy to use for large-scale research. Such data is carefully regulated to ensure patient privacy, and even if it were not, the logistics of gathering and sharing data among disparate medical organizations would still be challenging to say the least. Participants in the UK Biobank have given consent for their data to be added to a centralized database. Second, traditional medical tests are typically performed at unpredictable times, generally based on when patients develop symptoms or signs of disease. The prospective design of data gathering for the UK Biobank allows people to be tracked at regular intervals both before and after the onset of illness.

In a 2016 paper describing results from only the first five thousand participants to be imaged, Karla described how they had "already yielded a rich range of associations between brain imaging and other measures collected by the UK Biobank."[14] By 2022, Karla and her colleagues were reporting the sobering finding that COVID-19 infection is associated with changes in brain structure.[15] A growing number of research groups around the world are also taking advantage of the biobank's rich reserves of public imaging data. The broad aim of this collective work was well characterized by Karla in 2016: to use imaging to detect "a broad range of diseases at their earliest stages, as well as provide unique insight into disease mechanisms."

Imaging has become an important part of other population-level data collection initiatives as well. Historical studies like the Framingham Heart Study, begun in 1948, have expanded their purview to include cross-sectional imaging. More recent initiatives have included advanced imaging from the start. The NIH is building a research program called All of Us, with a scope similar to that of the UK Biobank. Big Tech has even dipped a toe into the medical image collection pool, with Google's Project Baseline.[16]

Karla may be a current maven of big imaging data, but she started small. Like Jürgen, and like me, she got her early training in the manipulation of tiny nuclear magnets. In some circles, the art of MR pulse sequence design is called spin gymnastics, since the signal we see actually results from a dizzying routine of microscopic flips and spins. (Karla is no stranger to gymnastics. Somewhere on YouTube, there is a video, set to guitar-heavy grunge music, of her as a teenager executing high-speed leaps on inline skates.) Karla's penchant for mental gymnastics took her from the University of Illinois Urbana–Champaign (where she studied computer science) to Stanford (where she received a master's degree and PhD in electrical engineering) and then to Oxford (where she has been based since 2004 as a fellow, lecturer, professor, and most recently research director).

She carried her interest in the brain with her from childhood. When Karla was twelve or thirteen years of age, her mother needed brain surgery, and young Karla found herself wondering whether the surgery might change who her mother was. She started out as a psychology major in college, until she saw functional MR images in a textbook and decided that they offered the best tools she had yet encountered for understanding how the brain works. (Are you detecting a common theme by now in how imagers are made? As often as not, they are drawn in unexpectedly by the power of images, or else they are attracted to the remarkable machines that make images.) Karla went on to develop expertise in a range of brain imaging tools. Her research group routinely uses diffusion imaging to probe microscopic brain structure, as well as fMRI to assess brain function and connectivity.

Today, Karla is a sought-after lecturer and a fixture at international imaging meetings. She is crystal clear as a speaker, drawing you in with her easygoing American accent peppered with hints of upper-crust Oxford English. She is quick with a smile but equally quick to call out an outdated point of view. If you try to get too easy a bead on her, you might find yourself caught between her distinguished head of silver-streaked hair and the eyebrow piercing she has sported for as long as I can remember.

Karla chaired the ISMRM annual meeting program committee during my tenure as ISMRM president. Working with Roberta Kravitz and her team, she coordinated the efforts of nearly a hundred expert volunteers to put together our 2018 meeting in Paris. In retrospect, it is clear to me that asking Karla to serve in this role was one of my very best decisions. She delivered a top-notch scientific program that also prioritized engagement of the full breadth of our imaging community. She designed "secret sessions" to demystify the profession of imaging and provide unvarnished advice on career and life hacks to young scientists. And she worked to ensure that the program committee membership reflected the current diversity of our field.

One of the only things more striking about Karla than her ferocious intelligence is her uncompromising compassion. She is committed to expanding the scope of imaging not only by visualizing thousands of brains with the UK Biobank but also by calling attention to the many faces and voices that make up the community of imagers. In addition to being her center's associate director, she is its diversity and inclusion champion. She is active on social media, highlighting the successes of underrepresented scientists and pointing out the ongoing biases they face. She does her best to be an antidote to preconceptions of who an imager can be (like the preconceptions of that cardiologist decades ago who failed to see Rod for who he was). I would argue that Karla is merely aiming to deliver on a promise implicit in the modern burgeoning of imaging technologies. As imaging expands to encompass our planet, it is only fitting for it to return to its roots as the property of everyone.

Had we but world enough and time, there are so many other people I would wish to introduce you to now. I would tell you about Dr. Hedvig (Hedi) Hricak—a one-time competition-class downhill skier from Croatia, former RSNA president, longtime chair of radiology at Memorial Sloan Kettering Cancer Center, and friend of all those who face cancer, whether as a caregiver or as a patient. I would spin you stories about the late great Graham Wiggins, a.k.a. "Dr. Didg," a master of the Aboriginal Australian didgeridoo who applied his formidable improvisational skills not only to sound waves but also to radio waves, designing out-of-the-box devices to listen for faint MRI signals. I would offer an appreciative nod to Professors Esther Warnert in Rotterdam and Udunna Anazodo in Montreal, who develop creative new uses of MRI and PET to study disorders like brain tumors, dementia, and depression, and who have worked together on creative new ways to combat bias in the imaging community. I would shine a light on Professor Shy Shoham, an Israeli-American neuroengineer who uses lasers to make ultrasound images, who creates holograms to decode the sense of smell, and who wants to change the way you think by dazzling your brain with patterns of sound and light. I would also boast about

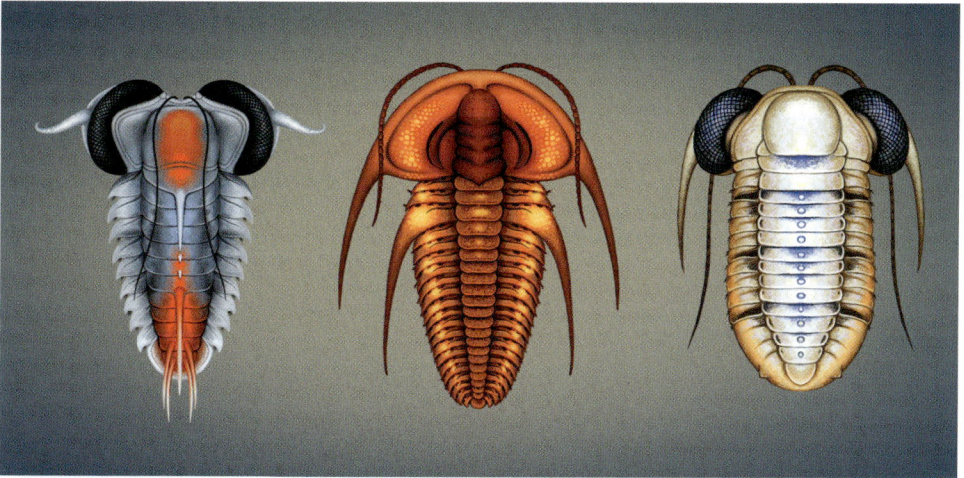

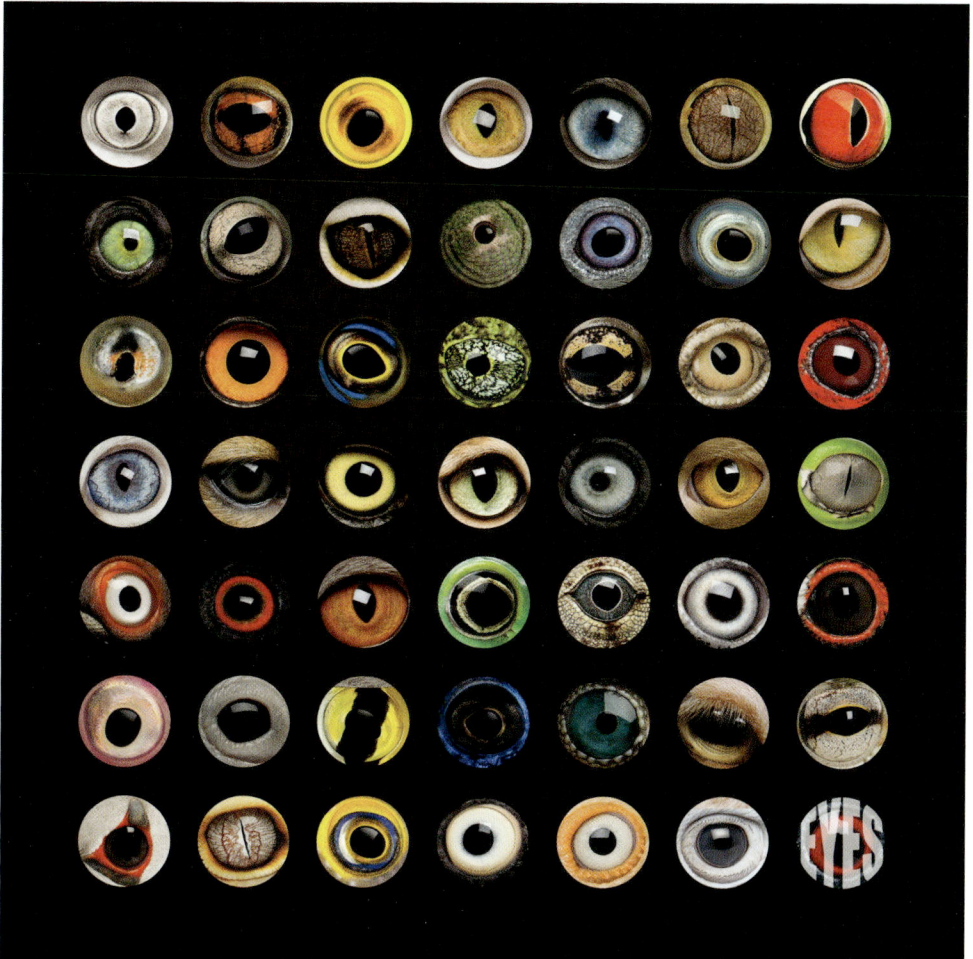

PLATE 1 The nature of seeing. (*Top*) Meet the trilobites, pioneers of vision. (*Bottom*) The startling diversity of animal eyes.

*Sources*: (*Top*) Courtesy of Franz Anthony (https://franzanth.com/); (*bottom*) Wanderlust 2003, CC BY-SA 4.0, via Wikimedia Commons.

PLATE 2 Augmenting nature and seeing it through. (*Top*) Handmade images. (*Middle*) Imaging with machines. (*Bottom*) Early x-ray imaging.

*Sources*: (*Top left*) Pablo A. Gimenez, CC BY-SA 2.0, via Wikimedia Commons; (*top center*) Marco Ossino/Shutterstock; (*top right, middle*) public domain via Wikimedia Commons; (*bottom*) 1985 painting by D. Jacques Rohr showing an x-ray fluoroscopic examination of a woman in 1896, courtesy of Centre Antoine Béclère, Paris.

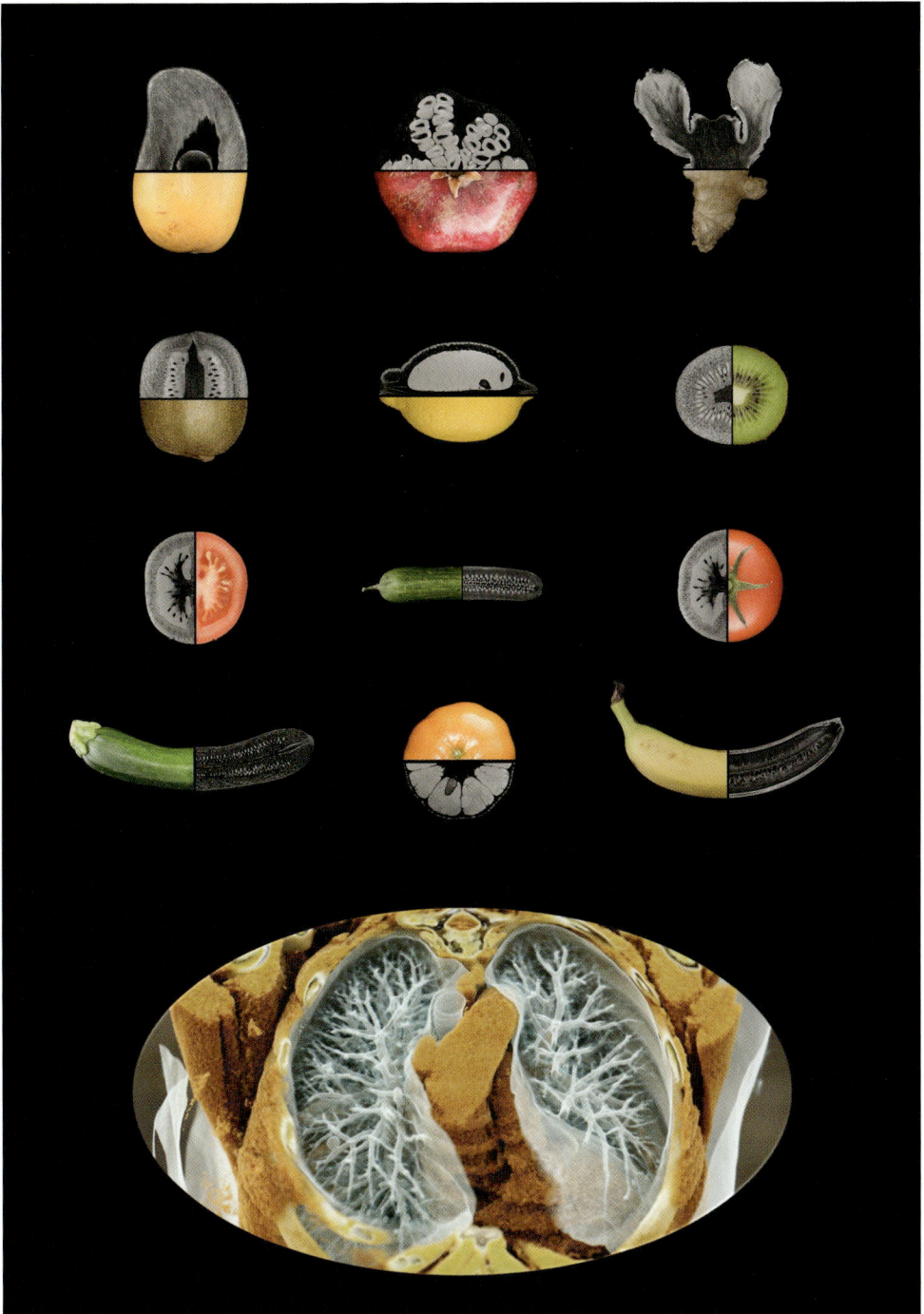

PLATE 3 Slicing without cutting. (*Top*) MRI still life. Slices of assorted fruits and vegetables spliced with color photographs. (*Bottom*) The body in living color. A cinematic rendering of lung anatomy from CT slices.

*Sources*: (*Top*) Courtesy of Pippa Storey and Pawel Slabiak; (*bottom*) courtesy of Siemens Medical Solutions USA, Inc. The CT data used to create these images were acquired at the Portuguese Institute of Oncology, Lisbon.

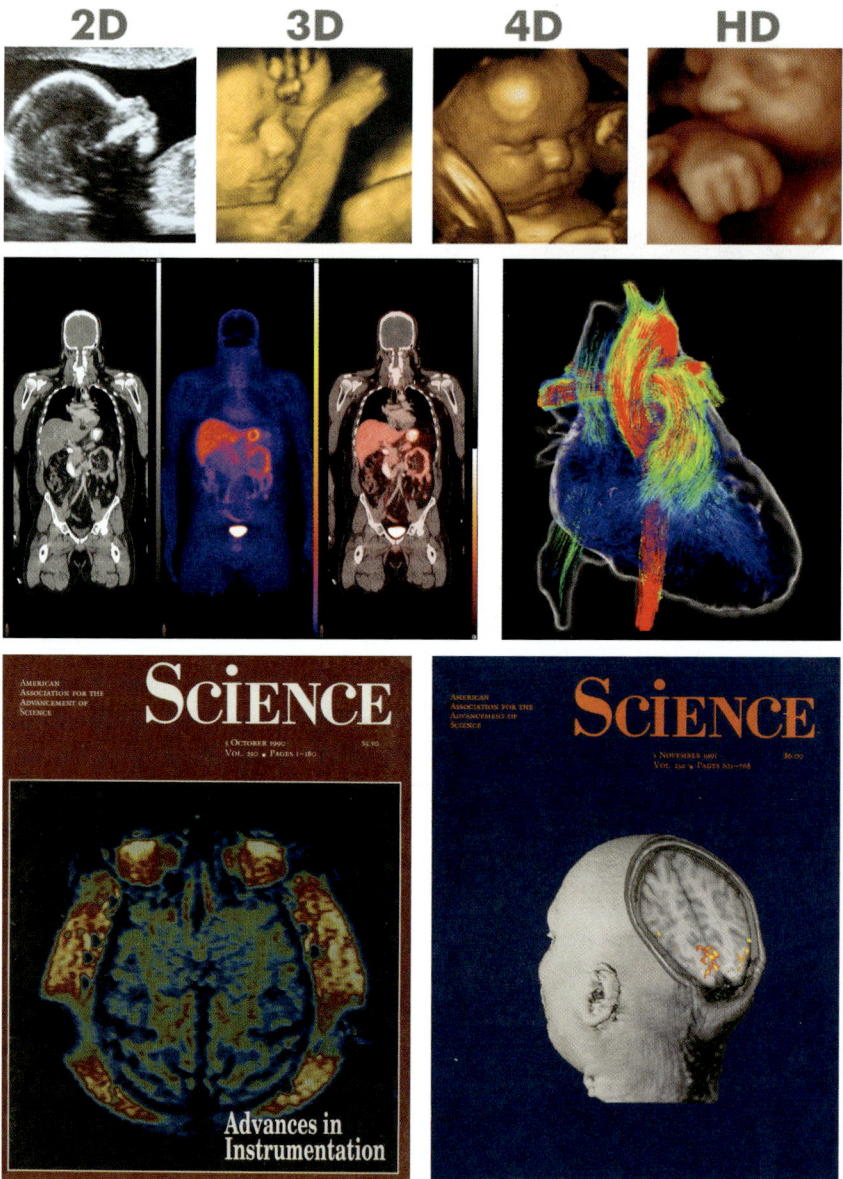

**PLATE 4** The many faces of tomography. (*Top*) Progressive improvements in prenatal ultrasound. (*Middle left*) MRI, PET, and fused MRI and PET images of the body. (*Middle right*) Tracking blood flow in the heart using MRI. (*Bottom*) Imaging advances featured on the cover of *Science* magazine in the early 1990s.

*Sources*: (*Top*) Public domain; (*middle left*) Department of Radiology, NYU Grossman School of Medicine; (*middle right*) courtesy of Rizwan Ahmad (https://u.osu.edu/ahmad/research/); (*bottom*) reprinted with permission from the American Association for the Advancement of Science; (*bottom left*) *Science* 250, no. 4977 (1990), visualization by Geoffrey Sobering, National Institutes of Health; (*bottom right*) *Science* 254, no. 5032 (1991), courtesy of the Massachusetts General Hospital NMR Center.

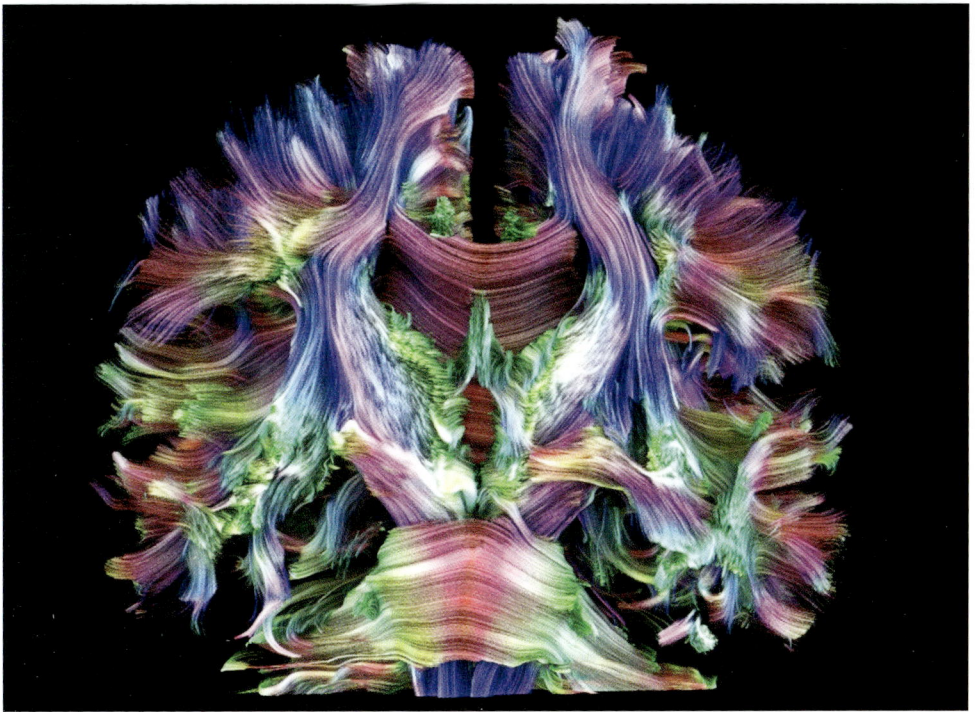

PLATE 5 What's in an image? Outer space and inner space. (*Top*) Harmony of the spheres. A composite image of the Crab Nebula in multiple frequencies. (*Bottom*) Symmetry of the mind. Tracing nerve fiber bundles in the brain using diffusion MRI.

*Sources*: (*Top*) Public domain via NASA (https://svs.gsfc.nasa.gov/30944). (*Bottom*) Courtesy of Dorin Comaniciu; data from the NYU Grossman School of Medicine; 3-D rendering by Siemens Healthineers.

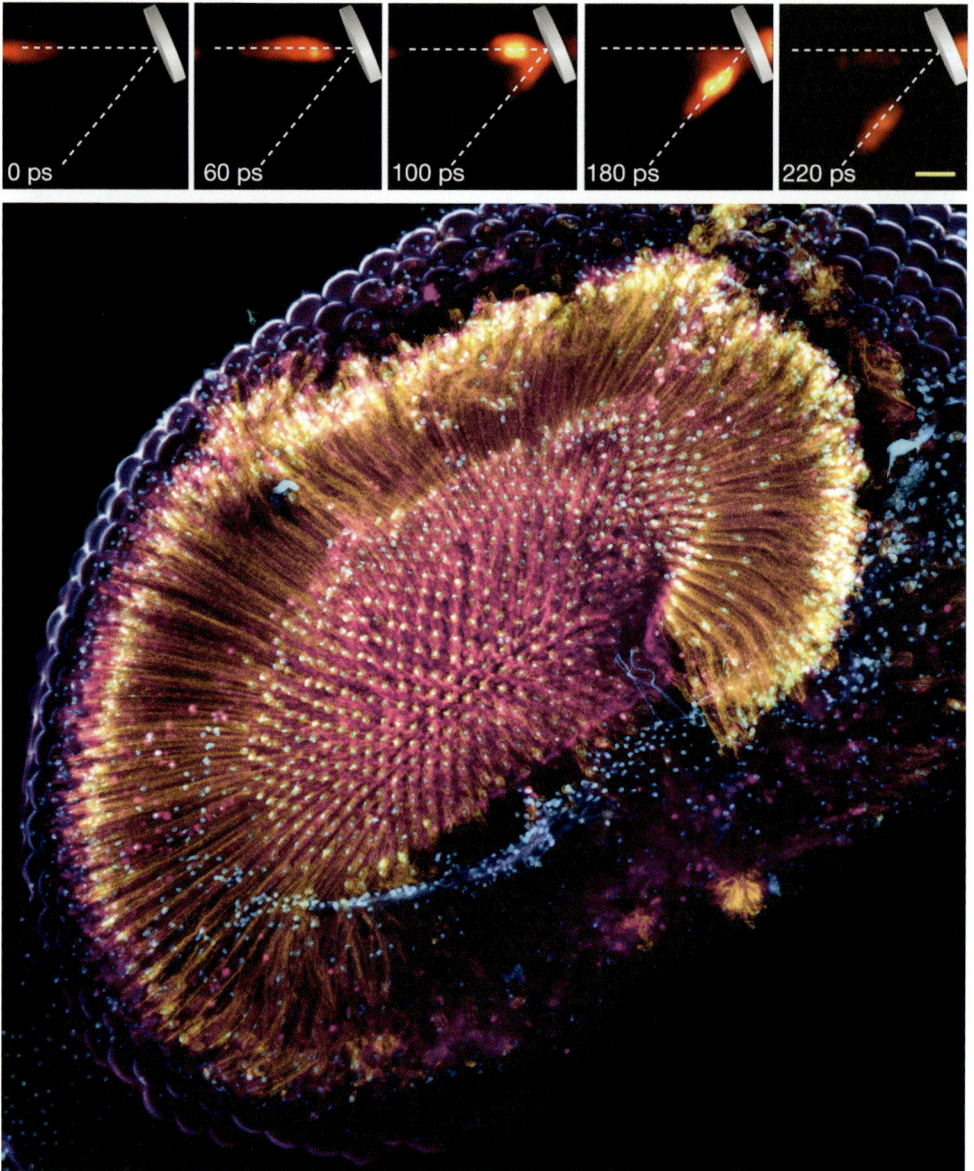

PLATE 6 "I have seized the light. I have arrested its flight." (*Top*) Compressed ultrafast photography depicting light bouncing off a mirror at 100 billion frames per second, delivering in the twenty-first century on Daguerre's exuberant nineteenth-century proclamation. (*Bottom*) A light microscope image of a fruit fly retina with its photoreceptors highlighted in yellow.

*Sources*: (*Top*) Reproduced with permission from L. Gao et al., "Single-Shot Compressed Ultrafast Photography at One Hundred Billion Frames per Second," *Nature* 516 (2014): 74–77, figure 3a. © Springer Nature; (*bottom*) Guillaume Thuery, CC BY-SA 4.0, via Wikimedia Commons.

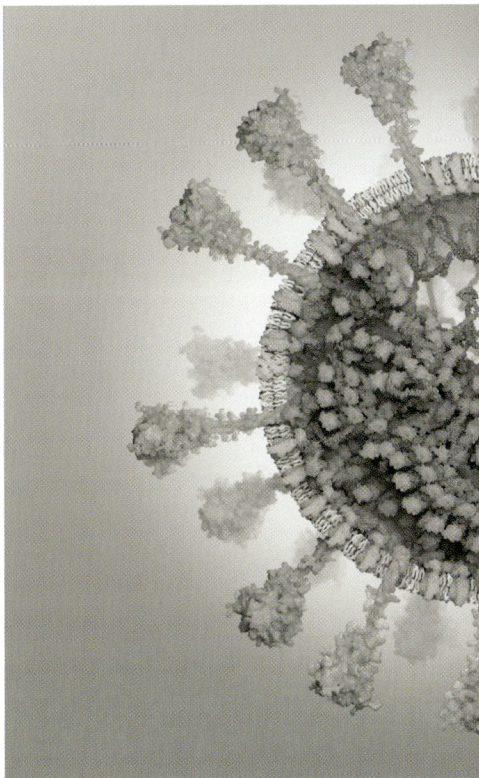

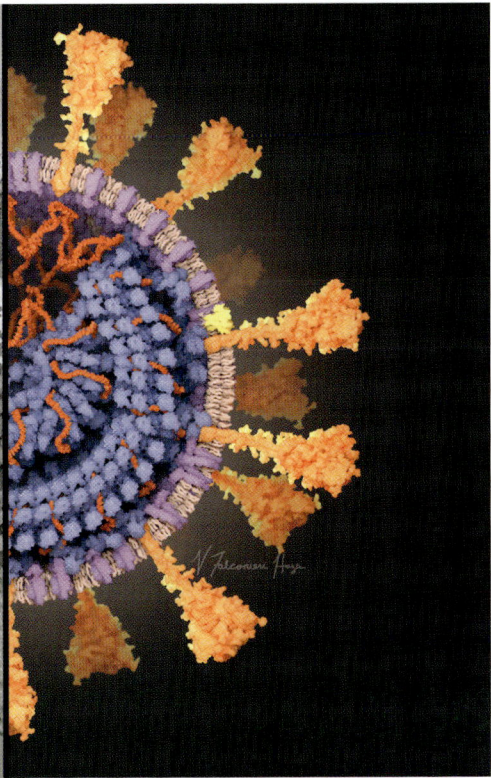

PLATE 7 Farther, smaller, clearer: imaging marches on. (*Top*) The Pillars of Creation as viewed by the Hubble (*left*) and Webb (*right*) space telescopes. (*Bottom*) A composite image of a SARS-CoV-2 virus, combining information from electron microscopy, cryo-electron microscopy, x-ray crystallography, and magnetic resonance.

*Sources*: (*Top*) NASA, ESA, CSA, STScI, Hubble Heritage Project (STScI, AURA); (*bottom*) © 2020 Veronica Falconieri Hays.

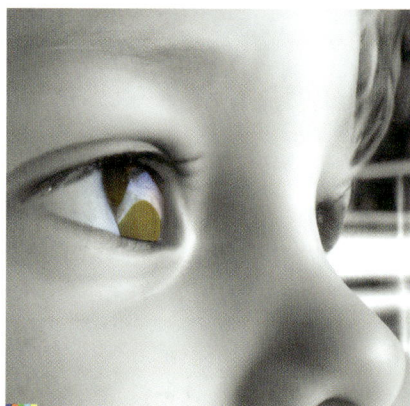

PLATE 8 The future of seeing. (*Top*) Bat eyes and ears. (*Bottom*) What does DALL-E see? The output of OpenAI's image-generating system following the prompts "A woman holding a miniature MRI machine in her hand" (*left*) and "The future of seeing" (*right*).

*Sources*: (*Top*) aaron007/iStock.com; (*bottom*) images generated by the author using DALL-E 2 (OpenAI).

my brother, Dr. Aaron Sodickson, a physicist turned physician, a leader in emergency radiology, and a CT innovator who has spent years tracking both deleterious effects and beneficial new uses of x-rays.

There are just too many people, and there is too little time to do any of them justice.

The experience of modern imaging writ large, of course, is much broader still. You've met a small sampling of stakeholders in this chapter: beneficiaries of medical imaging, and its practitioners; leaders of radiology departments and hospitals; civil servants responsible for funding imaging research; also researchers, who use that funding both to develop new imaging technologies and to find new uses for them; and captains of industry, too, charged with bringing the fruits of imaging research and development to market. Now imagine, if you will, a corresponding but distinctive collection of stakeholders in every other field of imaging. Consider the global enterprise of astronomy. Not only are there networks of earthbound and orbiting telescopes, but there are also worldwide networks of astronomers, engineers, technicians, and administrators dedicated to keeping them in good working order. Modern microscopes, manufactured by an array of modern companies, can be found in laboratories everywhere. Colossal industries have grown up around visual technologies designed not just for the professional specialist but for the interested consumer as well. The number of people engaged in the design and manufacture of cameras, video recorders, and other such devices has boomed, as has the number of people using these devices.

It is well known that species split, specialize, and diversify as they evolve. The same can be true of technologies and the communities that grow up around them. While artificial imaging may have begun with a few simple parlor tricks with lenses, it ended up encompassing a menagerie of brilliant variants worthy of evolutionary biology. Imaging in the modern world can be viewed as a great tree with many branches. This intricate phylogeny can be depicted in many ways. It can be sorted by imaging modality, the technology a particular group favors. Separate subpopulations of imagers now devote themselves to advancing the state of the art in MRI, CT, PET, ultrasound, x-ray, or optical imaging. The same is true for different wavelengths of light. Radio astronomers have their unique careers and day-to-day concerns, as do x-ray astronomers and visible-light microscopists. Another way imagers tend to sort themselves is by the specific target of their imaging devices. The search for exoplanets often has different devotees than does the investigation of exotic black holes. Medical imaging has its specialists in different body parts: brain imagers, musculoskeletal imagers, and imagers of the heart, to name a few. Some photographers specialize in portraiture, others in finding secrets in satellite imagery. Then there are the underlying interests that bring people to imaging. There are tool builders, who delight in finding new and improved ways to see. There are those who use imaging solely as a stepping-stone to discovery. Some are passionate about improving health, others about enriching local economies, or making art, or recording personal experiences for posterity. All these people make up the great living tree that imaging has become.

People today make their way to imaging from all walks of life. They span the world's nationalities, religions, gender identifications, and sexual orientations. They cover a broad spectrum of political opinion and personal taste. They are young and old; Black, White, and Brown; deaf and hearing; neurotypical and neurodivergent. Regardless of who you are, they look like you, and they think like you. Much work remains to be done, though, in ensuring that modern imaging is truly for everyone.

# 8

# Imaging for Everyone?

How disruptive forces and disruptive innovation are poised
to change imaging, and life, forever.

The story of imaging so far has been a story of expansion: a great
tree branching and blossoming, technologies and communities
burgeoning with increasingly global reach. The development of
imaging technology over time also serves as a case study in disruptive
innovation. Imaging has changed how many disciplines and industries
operate: medicine, science, news, entertainment, transportation, communications, the list goes on. Imaging has changed our human way of life.

In this, the last chapter of part I, I'd like to review where this track
record of disruption has left us today. First, let's look at the technologies.
Telescopes have gotten bigger—a lot bigger. They employ a wide range
of probes across the electromagnetic spectrum and beyond, but all variants have by and large been pulled upward in scale by the demands of
resolution and sensitivity. The distinct mechanisms of magnification in
microscopes have also proliferated, but the devices themselves have more
or less stayed the same size. Some differences in packaging may be called
for depending on whether one is using electrons, x-rays, or visible light,
but microscopes are typically still tabletop devices. And what about the
space between the microscopic and the galactic? What about the humble
camera? Cameras, as you have probably noticed, have gotten a whole heck
of a lot smaller.

We haven't spoken in much detail about ordinary cameras since chapter 2, but there is no disputing that modern cameras have come a long way since Daguerre. Cameras today are digital, high-resolution, and high-contrast. They operate at multiple wavelengths, ranging from the infrared to the visible to the ultraviolet. They include charge-coupled device (CCD) detector arrays with ever higher megapixel counts, and ever larger reserves of digital storage to accommodate the associated raw data. Multiple cameras are entrained together to allow adjustment of exposure and focus after the fact. Linked cameras can synthesize multiple points of view in a tomographic juggling act. There are new kinds of cameras as well: for example, light-detection-and-ranging (LIDAR) cameras, which track the time of flight of light beams for radar-like ranging or ultrasound-like three-dimensional imaging.

At least as much as for astronomy, microscopy, and medical imaging, the development of photography has been an exercise in disruptive innovation. Since the early days of Daguerre and his colleagues/competitors, photography has spawned a slew of new industries: camera companies, film companies, movie studios, and more. Then, advances in photography have overturned all those industries. People used to go to dedicated photo shops to develop film from their home cameras. Then they went to photo desks in drugstores. Then they stopped going anywhere at all when cameras went digital.

In addition to digitization, miniaturization has been a key disruptive trend. Originally conceived as tripod-mounted boxes modeled after darkened rooms, cameras today can be made smaller than our own eyes.[1] They are able to capture images without the need for consumable media like film, and they can transmit their digital images wirelessly. They can therefore hitch a ride on handheld devices like cell phones. They can be installed, less and less obtrusively, in all manner of consumer electronics and general infrastructure. They can be controlled remotely and maintained in operation using modest reserves of power.

As a result, cameras are everywhere today. Indeed, photography is perhaps the most obvious example of imaging for everyone. There are more cell phones now than there are people, and many of those phones are equipped with one or more digital cameras. In 2020, more than three billion images and seven hundred thousand hours of video were shared online every day.[2] The daily haul of shared images in 2025 is believed to exceed fourteen billion, and more than two trillion photos are taken worldwide each year.[3] We are utterly awash in images.

Consider for a moment the impact of this flood of images on just one sociological axis of modern living. On one hand, the concrete visual documentation of events promotes transparency. Cell phone cameras, dashcams, and body cams have all helped to shine a light on abuse by those in positions of power. The nine-minute George Floyd video served as powerful fuel for the Black Lives Matter movement. Poignant images have captured war crimes and unmasked societal inequities, becoming important parts of our social consciousness and collective conscience. On the other hand, the ability to capture life as it happens also poses unprecedented threats, not only to privacy but also to equity. Cameras are increasingly being used by governments for surveillance of the populations they govern. Facial recognition technology has become a lever of control by ruling elites. These uses of imaging propagate imbalances in power rather than curtailing them. Like any disruptive tool, imaging can clearly be used for good or for ill, depending on who controls it.

What will happen as cameras continue to get better, smaller, and easier to integrate into just about anything—our clothes, our glasses, or even our own eyes? What wonders and terrors will be possible when a cloud of buzzing mechanical insects can construct and transmit images? What will we do when the walls themselves have eyes? Science fiction novels are replete with scenarios like this, but in reality the foundations are being laid right now. The impact of omnipresent imaging is a theme we will explore further in part II.

In the meantime, let's check in on how other modalities are doing when it comes to universal access. Microscopes are widely available to hobbyists, to be sure, but state-of-the-art machines remain largely the province of experts. The same is true for telescopes. It is easy to procure a well-engineered backyard scope, but the leading-edge observatories are anything but accessible. Even if you're an astronomer, you need to put in competitive applications for observing time. When discoveries are made with the world's great telescopes, on the other hand, the resulting images rapidly go viral. Witness the global reception of the first images from the James Webb Space Telescope in 2022. These images are now part of the collective vision of our species. These products of imaging are indeed for everyone.[4] Likewise for raw astronomical data. Astronomy is well ahead of, say, the medical imaging field when it comes to data sharing. To be fair, data sharing might be a bit less of a heavy lift for astronomers than it is for medical imagers, since planets, stars, and galaxies tend to be less picky about their privacy than the human subjects of medical imaging. Nevertheless, astronomers deserve kudos for making broad access to their data a priority. When my son, Noah, developed an interest in astronomy as a teenager, he quickly figured out how to pull public data from the internet to search for planets around stars other than our sun. Astronomers seeking to justify the jaw-dropping resources required to put together new and complex astronomical instruments legitimately cite the contributions these instruments will make to human knowledge and the public interest. With its grand ambitions and international consortia, astronomy has long been a poster child for species-wide collaboration.

What about medical imaging? Putting aside the thorny question of access to medical data, a lack of universal access to imaging machines is more problematic for medical imaging than for astronomy or microscopy, since medical imaging could in principle benefit everyone not just collectively but individually. Indeed, if omnipresence is a hallmark of the future of imaging, then medical imaging devices have very much the opposite problem to that of cameras. They are not widespread enough.

You saw in chapter 7 that the modern medical imaging enterprise has a sprawling socioeconomic reach by many measures. It does not extend everywhere, however. Quite the contrary: it currently leaves vast gaps in space and time unimaged. In 2016, all of sub-Saharan Africa, with its population of 370 million, was served by only eighty-four MRI machines.[5] At that same time, the NYU School of Medicine's Department of Radiology alone operated nearly half that number of MRI units, and New York City was home to quite a few more machines than that. Even in the United States, where more than thirteen thousand MRI machines currently serve a population of nearly 330 million, most machines are concentrated in infrastructure-heavy regions, leaving large swaths of the country sparsely covered. The same is true for other countries well endowed with medical imaging devices. Overall, two-thirds of the world's people currently have no access to MRI.

Meanwhile, the way advanced imaging is currently performed restricts its use not only to specialized centers but also to selective situations. In standard medical practice, imaging is generally ordered as a follow-up to other medical findings. Patients typically present with symptoms, or abnormal lab test results, or other concerning indicators of significant risk, before physicians call in the "big guns" of cross-sectional medical imaging. Somehow, despite our common background as imaging creatures and our voracious appetite for images in many other walks of life, we have managed to box medical imaging in—to confine it to narrow tubes and to define it as the exclusive domain of experts.

This will change, and I maintain that the resulting changes will help usher in the next imaging revolution. Before that happens, though, medical imagers will need to address the elephant in the room: their bulky machines. For MRI, as you have learned, the demand for improved sensitivity and spatial resolution has led to bigger and bigger machines, which are expensive not only to purchase but also to install and maintain. While CT and PET are not subject to quite the same physical constraints, these machines by and large have gotten sleeker but not smaller over time.[6] In the normal course of development, new bells and whistles are added to medical scanners all the time, and neither physicians nor researchers like to settle for lower image quality than they have had access to before. Thus, for medical imaging as for astronomy, "bigger is better" has been the norm—at the cost of accessibility.[7]

Even in medical imaging, though, an opposing trend has quietly been gathering steam. In fact, a conviction that less is more is what first brought me to imaging.

My career in imaging got started in the usual way, which is to say completely by accident. Arguably, it was the result of an absent-minded doodle.

This was in the spring of 1996. As part of my medical training (which was also something of an accident, but that's another story), I found myself in a month-long rotation in the laboratory of a cardiac imaging expert named Dr. Warren Manning. My assignment was to write a report on a topic of interest related to Warren's area of study. When I first sat down in his office at the Beth Israel Hospital in Boston, Warren suggested that I look through some literature on MRI of the heart, then choose an area of focus that spoke to me. As I boned up on the basics of cardiac MRI, the problem of motion came up time and time again. The heart moved quickly in the chest, but MRI was slow. Without special measures, MR images of the heart would be blurry, like pictures of an athlete in action taken at a slow shutter speed. It would be very bad form to stop the heart in order to image it, so what other measures could be taken to clear things up? This was when I first learned about the trick of breaking up images across multiple heartbeats. It was an undeniably clever trick, but it was also cumbersome. Other techniques to track motion and correct it after the fact were similarly clever but cumbersome. Not knowing any better, I asked myself why we couldn't just image faster.

The shutter speed in MRI is set by the time it takes to gather a sufficient number of projections. Each projection takes time to read in, and adjusting magnetic field gradients to switch from one projection to another takes more time still. Back in 1996, high-resolution images typically took somewhere between seconds and minutes to acquire—far less time than had been required in the early days of MRI but still far too long to avoid blurring a fast-moving organ like the heart. MRI technology was advanced enough by then that dramatic improvements in shutter speed for each projection seemed like a heavy lift. But why, I found myself wondering, did projections have to be gathered one after the other? Was there any way to gather them in parallel? Could an MRI machine be made less like a scanner (which collects one bit of information after another) and more like the eye (which captures an entire scene all at once)?

At this point, I should note that by walking you through my early thinking here, I have no intention of overblowing my own role in the remarkable collective history of invention and reinvention that constitutes the story of medical imaging. Consider this diversion another quick gallery portrait designed to give you a little perspective on how your narrator came by his current views on imaging.

Enter the doodle. I had been reading an article about the use of multiple MRI detectors (called coils since they are generally constructed from loops of wire like old-fashioned TV antennas). Arrays of coils had been developed to improve signal-to-noise ratio over extended regions of the body. All coils gathered projections using the same configuration of magnetic fields at any given time, but each coil contributed signal from a different part of the body. Signals from the various coils were carefully combined, but they were not used in concert for purposes of spatial discrimination. One day, I found myself sketching—on an actual napkin if you can believe it—a new way to combine signals from different coils to generate new projections. If you could do this, then you could take a smaller number of traditional time-consuming projections, filling in what was missing after the fact using mathematical transformations. In other words, you could gather projections in parallel. If you managed to gather half your projections at the same time, then it would take you half the time to collect everything you needed for an image. MRI would suddenly be twice as fast.

I spent an uneasy, sleepless week trying to convince myself that the new parallel projections were real and that I wasn't peddling some kind of free lunch. When I was confident that they were, and that I wasn't, I went to Warren and asked him out of the blue if I could join his lab to work on the idea. To my surprise and eternal gratitude, he said yes.

For me, this was the beginning. I wrote my first imaging paper, and filed my first patent, describing the technique I had come to call simultaneous acquisition of spatial harmonics, or SMASH for short.[8] In 1997, I delivered my first conference presentation on SMASH at the ISMRM annual meeting in Vancouver. I was terrified. Sure, I had done my graduate work on magnetic resonance, but my areas of expertise were MR physics and molecular structure determination. Imaging was a different beast entirely, and I felt entirely exposed. I recall chewing through the better part of a bottleful of antacid tablets before my talk. Afterward, people gathered in the poster hall to debate what I had presented, and I got my first up-close-and-personal experience of the larger community of imagers. I was treated to a spectacle of scientific rigor warring with passionate opinions, insatiable curiosity offset by insistent skepticism. Some people told me they had already come up with the idea of parallel imaging years ago.[9] Others told me I was crazy. I was hooked.

Some of the people in that community were my colleagues. There was my boss and mentor, Warren, of course, who was equally adept at research, clinical care, and being an all-around mensch. Then there was Mark Griswold, who quickly recognized the potential of parallel imaging and became an early coauthor of mine at Beth Israel.[10] I actually owe my first oral presentation on SMASH to Mark: he yielded up to me a presentation spot he had earned at the ISMRM meeting for work we had done together, since my original SMASH presentation had been relegated to a poster (figure 8.1). I worked closely with Peter Jakob—a gruff and perceptive research fellow from Germany who became a brainstorming partner and gym buddy—and received guidance from Bob Edelman, an innovation-minded radiologist to whom MRI pulse sequences came so naturally that watching his fingers dance across the keyboard of an MRI console was like watching a concert pianist at work.

Other people became my competitors: Klaas Prüssmann and Markus Weiger, then graduate students at the Eidgenössische Technische Hochschule in Zurich, puzzled together over SMASH during a canoe trip

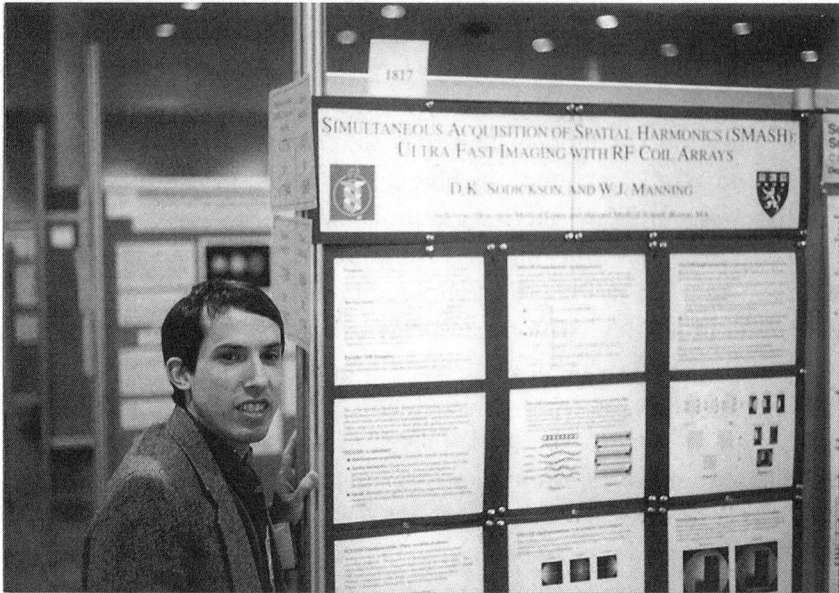

**FIGURE 8.1** Imaging in parallel. A portrait of the artist as a young man, proposing parallel MRI at the 1997 ISMRM annual meeting in Vancouver.

*Source*: Daniel K. Sodickson.

after the Vancouver conference and came up with a new parallel imaging technique they called SENSE.[11] For several years thereafter, SMASH and SENSE battled it out at conferences, and the competition fueled interest and experimentation. Over time, connections between various flavors of parallel imaging became clear, and people began to focus instead on applying it whenever speed was important. The MRI vendors got involved, and parallel imaging was introduced as a commercial product. The number of detector coils in typical MRI machines grew, paradoxically increasing the total quantity of raw data those machines generated, while at the same time decreasing imaging times. Parallel imaging was off to the races.

✧

MRI, of course, is not unique in its need for speed. No matter how you slice them, organs and bodies move. Even if you are interested in something comparatively stationary like an immobilized knee or head, people still don't like to hang out in big metal tubes while you take your time imaging them. Nor do radiology departments like to spend any more time than necessary on expensive imaging exams. At the same time that MRI was going parallel, in the late 1990s, a similar thing was happening in CT. With fast-rotating x-ray gantries, CT had already been substantially faster than MRI. When new CT machines came out that could gather multiple slices simultaneously in multiple rows of x-ray detectors, CT became blindingly fast. Ultrasound, meanwhile, had already beaten MRI and CT to the punch with highly parallel transducer arrays and a corresponding capacity for highly dynamic imaging.

I am telling you this story of parallelism not merely by way of self-introduction, and not merely to bring home the importance of imaging speed. Rather, I am highlighting for you some scattered seeds of the less is more movement in medical imaging. Parallel MRI gained speed by gathering fewer time-consuming projections than had previously been required for tomography and then synthesizing what was missing using transformations. It leaned on mathematics and comparatively lightweight detector hardware to do some of the job once assigned to heavy-duty scanner engineering.

This general trend has only accelerated over time. A decade after my first adventures in parallel imaging, another way of making images from limited sets of projections appeared on the scene. This method was grounded in a new set of mathematical transformations. It was called compressed sensing.

You can think of compressed sensing as a kind of image compression, but in advance. One way or another, most of us are familiar with image compression. For example, when we want to send a photo we've snapped on our phone to a friend, we may be asked whether we prefer a small, medium, or large image. Image compression algorithms like the popular JPEG algorithm analyze images and throw away parts that don't add important detail, trimming down the size of the files needed to store the images. Like natural images, medical images are also compressible, sometimes significantly so. So, why do we spend precious time in tomography gathering full sets of projections if we know that we can throw much of the information away after the fact?

Compressed sensing tells us that we really don't need to gather everything in advance. As long as we gather projections in the right way and use the right algorithms to reconstruct images, we can get away with significantly fewer projections. Several mathematicians of note made their name in the early 2000s developing such algorithms and demonstrating their value for numerous applications ranging from digital communications to photography. They also showed that the best way to gather projections was in a random-appearing or "incoherent" pattern. It was a graduate student at Stanford named Michael (Miki) Lustig—now a colleague I admire and a congenial fellow traveler in imaging circles—who introduced compressed sensing into MRI in 2006.[12] It could be combined with parallel imaging for additional gains in speed, but it didn't need multiple detectors to do its work.

Compressed sensing took off like a rocket. It found uses across the spectrum of medical imaging. Related approaches made serious headway in radio astronomy, a field in which computational scientists like Urvashi Rau Venkata (whom you met in Chapter 6) were struggling to reconstruct crisp images from limited data. In 2014, researchers using compressed ultrafast photography captured movies of a laser pulse bouncing off a mirror at the blistering frame rate of a hundred billion frames per second.[13]

The new, rapidly evolving field of computational optics leverages compression, parallelism, and other assorted tools and transformations to give cameras brand new capabilities. New tricks of perspective are being dreamed up all the time by optics researchers around the world—like Professor Laura Waller, whose laboratory at the University of California, Berkeley, is across the hall from Miki Lustig's. Laura and her team are making high-resolution cameras without lenses. They are taking a page from radio astronomy and MRI, synthesizing high-powered microscopes by coordinating less-powerful components. With heavy lifting done by math rather than by raw technology alone, ordinary cameras are proving to be capable of some pretty extraordinary things.

When it comes to computational imaging, the disruption is actually just getting warmed up. I have yet to tell you about one more horse that has recently entered the "less is more" race. I am referring to artificial intelligence (AI).

It is difficult to escape the reach of AI nowadays. Reports on advances in AI are splashed across the news media. Both the scientific and the popular literature are teeming with commentaries on its implications for the future of work and the nature of life as we know it. With help from Big Tech and little start-ups, in industry and academia and government alike, AI is making inroads into a terrifically (and terrifyingly) diverse set of human endeavors. Let me tell you just a bit about what it's doing for imaging.

Artificial neural networks are a key enabling technology at the core of the AI explosion. These networks are software constructs that take some cues from the organization of neurons in biological brains. Information flows in and is shuttled from one layer of neurons to another, undergoing a series of comparatively simple transformations that, taken together, can accomplish all kinds of complex tasks. Precisely which tasks they perform depends on how they have been trained: how the internal transformations have been adjusted to optimize their combined performance. Modern neural networks can play chess, or Go, better than the best human experts. They can parse complicated prompts and generate volumes of utterly realistic-seeming text in a dizzying variety of styles. They can write songs, emulate individual voices, streamline business processes, predict the weather, and solve previously unsolved problems in synthetic chemistry. The capabilities of modern neural networks have been amped up by new computational power, for example, in the form of graphics processing units (GPUs) which were originally developed to feed humanity's appetite for computer games. They have also been empowered by the availability of large sets of sample data that can be used for training. In the arena of computer vision, neural networks can now be trained using public databases of millions of images culled from the internet. They can tell the difference between cats and dogs and apples and oranges in online photos. They can generate their own photos, which are increasingly difficult to distinguish from the products of actual imaging. They can identify obstacles in real time for heads-up displays or self-driving cars. They can learn to detect tumors in medical images.

How can any of this be characterized as "less is more"? What I have described so far is all about bigger datasets, better computers, and more information. Nevertheless, around 2016, another ten years after compressed sensing made its entrance into imaging, people started using artificial neural networks to generate images from even more limited sets of projections.

The key here was the capacity of neural networks to learn. When you try to push imaging speed too much with compressed sensing approaches, you run up against the limits of fixed transformations like the JPEG algorithm. Images start to look blurry, or just unnatural. Subtle details disappear. AI, on the other hand, can learn which transformations work best to generate clear images from limited data. They can learn what crisp images should look like and try their best to generate new images that look that way.

At this point, you may find yourself objecting that the *last* thing you want to do is rely on learned assumptions about what an image should look like. This is particularly true in medical imaging, where seeing *unexpected* structures that might indicate disease is the whole point. What if you were to fill in the normal anatomy of, say, someone's liver and miss a tumor lurking there? Wouldn't that be bad? You are absolutely correct. However, researchers have devised a number of interesting ways to prevent neural networks from just seeing what they expect to see.

In his best-selling book, *Thinking, Fast and Slow*, the celebrated psychologist Daniel Kahneman describes two distinct "systems" built into our way of thinking.[14] System 1 is what drives us to leap to conclusions based on learned expectations and intrinsic biases. System 2 is our reality check: slow, deliberate, and rational, performing rigorous calculations and assessing the correctness of our conclusions. Well, it turns out that artificial neural networks can also be configured on this general model. Many of the networks now used to reconstruct images using limited projections include a component that learns from sample data and another component that keeps track of the known physics of tomography, forming projections of any candidate image and checking them for consistency against the projections that have actually been measured. In some recent studies, it has been shown that radiologists cannot tell the difference, in either diagnostic information or overall appearance, between fast images generated with this two-system approach and traditional slow images.[15] There are still limits to the imaging speed that can be achieved in practice—either SNR runs out or spatial resolution starts to fail if you push things too far—but those speed limits are higher than they were before.

The bottom line is that AI offers a new set of adaptable transformations that can help us to make more with less data. These transformations are loosely modeled on some of the rudimentary functioning of biological brains. Our brains, it should be noted, have also evolved to act quickly on data that may be incomplete. Parallel imaging and compressed sensing showed us that we could get away with less time-consuming imaging data than we once thought was necessary, without losing key image information. AI allowed us to do still more with even less.

AI is usually presented as a powerful way of drawing new conclusions from existing data. It is important to remember, though, that it can also allow us to use *different* data from what we once might have thought to use. In the context of accelerated medical imaging, the progression of transformations from parallel processing to image compression to neural networks has changed the way we are able to gather imaging data. We are no longer restricted to a simple march of ordered projections. Our data can be more disordered, more diverse, and sparser, trimmed down to the essential outlines. Advanced cross-sectional imaging technology was originally designed to build up images piece by piece, and for a long time imaging devices themselves advanced in much the same additive fashion. Now, with help from new transformations, imaging is in the process of deconstructing itself.

I've spoken about the role of transformations for less-is-more imaging. Now let's look at whether imaging technology can follow suit. Even if we require fewer projections than once we did, we still need a way of making those projections, and making them cleanly. In medical imaging, that has traditionally meant bulky tubes.

I am pleased to report that a less-is-more movement is currently gaining steam in the engineering of medical imaging devices (figure 8.2).

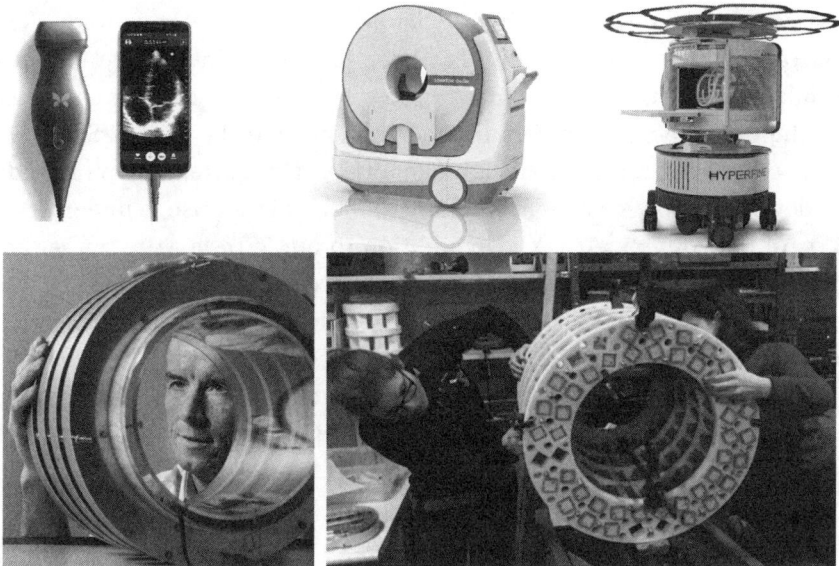

FIGURE 8.2 Less is more. (*Top*) Recent commercial offerings for handheld ultrasound (*left*), portable CT (*center*), and mobile MRI (*right*). (*Bottom*) Prototype low-field MRI systems currently under development by academic researchers.

*Sources*: (*Top left*) Courtesy of Butterfly Network, Inc.; (*top center*) courtesy of Siemens Medical Solutions USA, Inc.; (*top right*) © 2024 Hyperfine, Inc. Used with permission; (*bottom left*) courtesy of Andrew Webb; (*bottom right*) courtesy of Clarissa Cooley and Charlotte Sappo.

Ultrasound got the ball rolling some time ago. Gone are the immersive water tanks that were used with early human subjects. In their place are handheld ultrasound wands connected to rolling carts with computer displays. Recent inventors have gone further, replacing the carts with smartphones and making the ultrasound wand effectively a peripheral device for your phone. Ultrasound is a bit of a special case since you don't need to encircle the body to get a strong and sufficient signal. A single small wand can emit high-powered ultrasound pulses and record reflected echoes. CT, PET, and MRI are harder to miniaturize, but this hasn't stopped a growing community of entrepreneurs and scientists from trying. Concerted efforts are now underway in both academia and industry to produce cheap, mobile, or even portable CT and PET scanners. These stripped-down devices will be able to minister to people in a broader range of settings than their older, bigger, and higher-maintenance siblings.

In the field of MRI, the historical move toward higher magnetic field strength has been joined, and perhaps even outpaced, by a new race to the bottom. Many of my esteemed colleagues who made their names pushing the limits of high magnetic field strengths are now setting the bar as low as they can go. Other rising stars have launched their careers with low-field innovations. A new consortium has been formed to investigate low-field mobile brain imaging for children in low- and middle-income countries. For the first time, researchers in Europe, Asia, and the United States are joining forces with imagers in sub-Saharan Africa. Entrepreneurs in India are developing low-cost, lightweight MRI machines for use in underserved parts of the country. At NYU, in addition to our 7 Tesla MR scanner and various 3 and 1.5 Tesla machines, we now have a 0.55 Tesla scanner developed by Siemens and a 0.064 Tesla machine manufactured by a young company called Hyperfine. The Hyperfine machine plugs into an ordinary wall socket and is not even tube-shaped: a patient's head fits between two circular plates. While not exactly a featherweight, it is compact enough to be driven around on wheels or carried easily from place to place in a van. Its magnetic field is weaker than what Lauterbur used in the first MRI machine he built to make human images in the late 1970s. Yet another more recent addition to the NYU low-field MRI portfolio weighs in lower still at 0.043 Tesla. In October 2023, colleagues of mine organized a hackathon called MRI4ALL, in which participants from around the world gathered in New York to build a tabletop MRI machine from scratch in just four days. The first image they obtained emulated Lauterbur's original image of two test tubes. They then took a page from Hounsfield and imaged the contents of a mystery box.[16]

I don't want to give you the wrong impression. The images from these low-field systems are certainly lower in quality than traditional high-field images. They are much better than many skeptics thought they would be, though. This is in part because low-field scanners don't have to contend with certain practical challenges and engineering imperfections that go along with high magnetic field strength. If you no longer take it for granted that bigger is better, it turns out that you can leave some historical baggage behind. You can also shift some of the heavy lifting in image generation from expensive hardware to inexpensive software.

There is another important caveat to keep in mind. If one wants to use images from the new stripped-down imaging systems in the same way that state-of-the-art images have come to be used, expensive expert interpretation is still needed. One could even argue that specialty training becomes all the more essential as images get worse. Even with state-of-the-art handheld ultrasound, it is still quite difficult to understand what you are seeing—unless transformations can be developed to help with that, too.

Of course, one could also argue that any imaging is better than no imaging at all. One doesn't need top image quality or years of specialty training to identify many important disease processes that affect people in regions of the world currently underserved by imaging. In other words, a single model of how imaging should work does not necessarily apply in all situations. Perhaps it's time to start thinking outside the box.

In its need for speed and its drive for accessibility, medical imaging is finally beginning to emulate photography, not to mention other disruptive technologies that have changed our lives. Indeed, the general process of disruption and dissemination that photography has experienced has played out many other times in human history. Tasks once confined to humans were taken over by machines. The output of those machines was digitized. Digitization enabled coordination among ever larger collections of ever smaller machines. This has certainly been true for computing. It has also held true for communications. Alexander Graham Bell told Mr. Watson to "come here" from the next room over; before you knew it, we had world-spanning wireless communication networks that no longer needed voices at all. The world itself is not even big enough to contain these networks anymore. At this moment, Earth is orbited by many thousands of increasingly interlinked satellites. This artificial halo of devices communicates with the devices in our pockets, cars, and homes, allowing us to pinpoint exactly where any of us is at any given time and enabling spying eyes to zoom in on any spot around the globe. A similar trajectory of disruption may be discerned in astronomy. As you have learned, arrays of telescopes are routinely linked across large distances to emulate much larger telescopes.

This process of progressive coordination also emulates our own biology as creatures of imaging. Our compact eyes work in tandem to enable depth perception. Our highly distributed sensory nervous system coordinates many kinds of sensors to create for us a multifaceted picture of the world. Other living creatures have evolved to see the world in strikingly different, but similarly coordinated, ways. How might artificial imaging follow suit? This is what we will explore in part II.

Chapter 9 will examine new ways to gather imaging data continuously, as our eyes and other sensory organs do. In chapter 10, we will revisit AI, pondering whether we can emulate some of the human brain's more advanced functions not merely to interpret imaging data but also to change it. Chapter 11 will explore how medical imaging technologies might be linked across space and time to create a new safety net for global and individual health. We will close in chapter 12 by considering how a broad democratization of imaging might play out, and how it might change the way we see ourselves and the world around us.

In addition to being a wellspring of disruptive innovation, imaging is itself being disrupted by multifarious external forces. The ongoing revolution in camera design has been fueled by the ascendency of social media— and by the profit motives of Big Tech companies. When it comes to outer space and inner space, powerful external forces are pushing imaging in different directions in different parts of the world. Somewhat paradoxically, the dominant forces in resource-heavy regions today are as often as not inhibitory. The construction and operation of large-scale astronomical observatories have always been subject to budgetary pressures, as well as cycles of public interest and political will. Significant economic pressures also underlie the need for speed in medical imaging. Medical centers push to minimize the duration of imaging examinations so that they can best serve their patients, and also so that they can image more patients.

Meanwhile, even as the volume of medical imaging data has ballooned, with a typical radiologist being called upon to cast an eye over billions of pixels every day, the pressure for cost containment has caused reimbursement for imaging services in some imaging-heavy regions of the world to plummet. Advances in imaging speed alone have not been enough to keep pace. Some imaging practices have gone out of business. Others have begun to use abbreviated imaging protocols, stripping out everything but the data required to answer very specific clinical questions. In short, pressure is mounting to image less. At the same time, reaching everyone who might benefit from medical imaging calls for imaging quite a bit more. In these ways and others, imaging is at a crossroads.

Which way will the day-to-day practice of imaging go in the future? What new technologies will tip the balance of costs and benefits? What new transformations, their hour come round at last, will wreak havoc on our accustomed way of doing things, causing us to see things in a new light?

Both history and biology lead to one inescapable conclusion: neither "bigger is better" nor "less is more" is likely to be the answer in the end. The question is not merely whether to image more or to image less. The key will be to image *differently*.

# PART II

LOOKING FORWARD

The Future of Imaging

# 9

## Emulating the Senses

### *From Snapshots to Streaming*

How imaging in the future will be less like still photography
and more like a symphony of sensation.

Y ou are lying alone in a dark tunnel. Snug and warm around your
midsection, cooler where your head is exposed, you are both comfortably ensconced and, at the same time, strangely ill at ease. Loud,
unfamiliar sounds jangle your senses. You, my friend, are in the grips of
modern imaging.

Knowing what you now know about the history of imaging, this tableau
presents a picture of stark contrast. On the one hand, it is the culmina-
tion of millennia of innovation. Behold the Big Imaging Machine, with its
imposing tube and its banks of supporting electronics connected by thick
umbilical wires. Witness the orderly progression of projections that surround you from all sides, in order to reveal you on the inside. When a sleek
modern computer is done rendering these projections into digital slices,
the ordered stacks of slices are whisked off to the watchful, waiting eyes of
an imaging expert. It is all very engineered, very professional.

On the other hand, here you are. As the imaging process proceeds,
you are in the middle of it all, seeing, hearing, and feeling everything. You
might peer out of the tube into the room beyond, plotting your eventual
escape. You might startle a little when you hear the first thud or whine of
the machine beginning to do its thing. You might feel fear, or curiosity, and
this might focus your attention and color your perceptions. Your experience of being imaged is not orderly or mechanical. Despite all the nice

analogies we've explored so far, the imaging you are doing with your own faculties is, in practice, nothing like what the machine is doing. Neat stacks of spatially organized information are not delivered to your senses. What you experience is a festival of sensation.

There is of course no reason we should expect our machines to do things the same way that our bodies do. After all, modern imaging machines can do things our bodies were never meant to do. Why would we imagine that the actual practice of medical imaging should resemble ordinary eyesight, or that modern astronomy should look like ancient stargazing? That said, there are also things our bodies can do that leave our fancy machines in the dust. To forget this as we revel in technological glory is to miss out on important new opportunities to evolve.

Chapter 1 began with the natural evolution of senses like vision. Then we left the senses behind in the early oceans and explored the imaging tools humans learned to build once we had colonized solid ground. Now it's time to come back. It's time to return to our senses.

Our modern world is an always-on world. In our current information age, information—visual and otherwise—flows continually along ever-swelling and ever-diversifying channels. We are surrounded, in the words of Paul Simon, by "staccato signals of constant information."[1]

In the field of imaging, this trend is most immediately evident in the changing pace of photography. The art and science of photography have progressed over time from meticulously choreographed still frames to rapid-fire snapshots to real-time streaming video. Nowadays, ballooning video streams clog the bandwidth of our wireless networks and consume more and more of our collective attention. Other forms of optical imaging such as microscopy have followed suit, tracking living processes in real time with ever shorter shutter speeds and generating mountains of digital data along the way.

The embrace of video in medical imaging has been far more tentative. Many of our medical imaging protocols still hark back to the old days of x-ray exposures and photographic film. Patients are called upon to act like the models of old-fashioned photo shoots. They are asked to hold still, to hold their breath, or otherwise to suspend animation while the shutter is open. Only after all the exposures are done can everyone return to life as it really is. This will change.

Before delving into some of the changes that are in store, let us first take a moment to consider the origins of the current status quo. Why, after all the rich development summarized in part I, are you still instructed to pose for your pictures when you enter a radiology suite? Why is enforced stillness still the way of things in the third decade of the twenty-first century when so many other facets of life appear to be in constant motion?

Why? Because cross-sectional imaging is still painstakingly slow. Unlike for modern digital cameras, the shutter speed for many medical imaging devices is measured in seconds or minutes rather than in microseconds.

And why is the shutter speed so slow? Basic principles of tomography give us the answer. We must gather multiple distinct and carefully coordinated projections before we can stitch them together into cross-sectional images. The longer each projection takes, and the greater the number of projections required, the slower imaging becomes.

In chapter 8, you learned that things are getting better. Scientists and engineers have learned how to make rapid transitions between projections. They have figured out how to take multiple projections at once (as in parallel MRI and multidetector-row CT), rather than waiting for them to flow in one at a time. They have even learned how to take fewer projections overall while still generating sharp images (using compressed sensing, then AI). Each of these advances increased the effective shutter speed of medical imaging devices.

None of these advances, however, has succeeded in changing the underlying imaging paradigm, because shutter speed alone is not the whole story. For CT, MRI, and PET, heavily engineered tubes are still called for to gather projections that are sufficiently precise for state-of-the art imaging, regardless of the number of projections required. And once you're bundling people into tubes, other logistical considerations come into play. Consider MRI. A modern MRI protocol typically involves not just individual slices but stacks of slices covering extended volumes of the body—and one stack is seldom enough. Each image stack may be repeated multiple times with different underlying contrast or with the injection of specific contrast agents. Each individual acquisition can be accelerated with parallel imaging, yes, perhaps with a touch of compressed sensing here or a dab of AI there, but the whole kit and caboodle still takes time. The patient waits in isolation, while the operator in the control room outside scurries to manage a sometimes dauntingly complex workflow.[2]

We are leaving all kinds of information on the table in such complex protocols—pretty much literally. The scanner is idle in all the gaps between distinct image stacks and contrasts. It is also typically idle while contrast agent injections are being prepared and while patients on the table are being instructed to lie still, hold their breath, or otherwise prepare for scanning. A study published in 2010 showed that more than 40 percent of the time patients referred for abdominal imaging spent on the MRI table was "non-value-added time."[3] In other words, nearly half the time occupied by abdominal imaging examinations involved no imaging at all. In modern radiology, time is money, and time is information. So, why do we not acquire image data continuously as long as the patient is in the scanner?

There used to be a reason. Tomographic imaging cut its teeth in an era when electronic computing was new and difficult. Just getting a clear image out of an imaging machine and its onboard computer was a victory. As a result, an ethic of careful preparation developed. You didn't just gather data willy-nilly. Every bit of data gathered would have to be stored somewhere for later processing, and electronic storage space was precious. It made no sense to collect data that might be incorrectly calibrated or corrupted by motion since correction algorithms were crude, and bad data was a waste of space and time.[4] But in the modern world, storage is ample and processing power abundant. So, why do we continue to pause between MRI exposures?

The answer can be traced to a hidden set of assumptions. The old medical imaging paradigm is based on the time-honored concept of an image as a freeze-frame: one static picture of a world that is always in motion. In old-fashioned cameras, a shutter clicked open and shut, the film behind it was exposed, and an image was born. Modern smartphone cameras have dispensed with the film and the mechanical shutter, but they still allow you to specify the sound of a shutter click so that you know when a picture has been taken. This mechanistic instinct carries over to medical imaging, even though the exposures are quite different and are necessarily longer. You begin acquiring data at one point in time, and you finish when the image frame is done. You can take as many snapshots as you like to characterize a moving subject. You can string these snapshots together into a movie if you wish. But the fundamental challenge is to collect a crisply defined image frame.

Medical imagers have spent so much effort rising to that challenge that they risk getting lost in the underlying assumptions. They risk submitting to the tyranny of the image frame.

A quiet rebellion is already underway. Iconoclastic imaging researchers have recognized that projections don't necessarily need to be lined up just so. They have figured out ways to share information between image frames. They have even explored dispensing with strict notions of a pre-defined image frame.

Here is one example. A decade or so ago, one brilliant band of rebels I had the good fortune to work with at NYU developed a continuous imaging approach called GRASP.[5] GRASP was designed to allow new freedom of movement during MRI scans. Rather than proceeding row by row in orderly regiments—an approach that falters when patients don't follow their orders to stay still—GRASP projections jump around, each one filling the biggest remaining gap in a pattern that never repeats itself. This pattern works very nicely with acceleration techniques like compressed sensing and parallel imaging, which means that you can generate high-quality images with a comparatively small number of projections. More importantly, no single projection occupies a privileged position at the start or end of an image frame. There is in fact no fixed start or end time for any frame. You are free to choose which projections you group together to form each image, since they all cover angles that are complementary to one another. You gather projections in a continuous stream for as long as you choose, and you decide how to sort things out after the fact. Did your reconstructed image miss the exact time an injected contrast agent reached the liver? No worries. You just move a little forward or backward in the stream and try again.

You don't even need to stick to projections that were taken right after one another. If you are imaging the heart, you can use the time-tested technique of taking different bits of an image from different heartbeats, grouping projections depending on where the heart was when they were gathered. The same approach works for breathing motion, too. In general, you can sort the incoming stream of projections into different bins based on how patients or their organs are moving, in order to generate crisp sequences of images that capture that motion. Gathering lots of ever-shifting data and sorting it in this way even helps you to get around the shutter speed limits associated with traditional frame-by-frame image acquisitions. This is because body structures move in a coordinated fashion, and you can use information from one time to fill in gaps in your knowledge at another time. Contrary to prevailing assumptions, you don't need to treat time as a runaway train you are always racing to control. Instead, you can go with the flow, building up a clear picture as time passes.

This kind of freewheeling approach is also evident in another continuous imaging technique developed by a research group at the Cedars-Sinai Medical Center in Los Angeles. This technique is called magnetic resonance multitasking.[6] As the name suggests, MR multitasking accomplishes a number of things at once. It builds up volumes of image information to track moving organs like the heart, while at the same time measuring rates of signal decay to assess the health of each part of the moving tissue. This is a bit like playing a game of darts while riding a roller coaster. It's not about how fast you can throw; it's about how well you can adapt to changing conditions.

Magnetic resonance fingerprinting—yet another mold-breaking imaging approach, this one developed at Case Western Reserve University in Cleveland—introduces changing conditions even when the imaging target is motionless (figure 9.1, top left).[7] It varies not just the angle but also the timing and the strength of projections, mixing different contrast mechanisms together in a way that would confound traditional methods of image reconstruction. The result is a set of complex signal patterns that are quite distinctive but difficult to interpret—just like actual fingerprints. Taking inspiration from fingerprint databases, which associate distinctive patterns of whorls and curlicues with individual people, MR fingerprinting creates a database of signal patterns resulting from particular tissues and finds the best match for each measured signal. It arrives at detailed maps of tissue properties not by marshaling regular ranks of projections but by embracing complexity and learning to match the resulting complex patterns.

Each of the MRI examples I have just mentioned—continuous imaging, multitasking, and fingerprinting—abandons time-honored approaches to the collection of tomographic projections. Instead of compulsively repeating fixed patterns of projections, each gathers imaging data in an ongoing, mad rush, then sorts all the details out later. These methods, and a growing number of related cross-sectional imaging techniques, are united in a revolt against the tyranny of the image frame. In the place of that rigid old

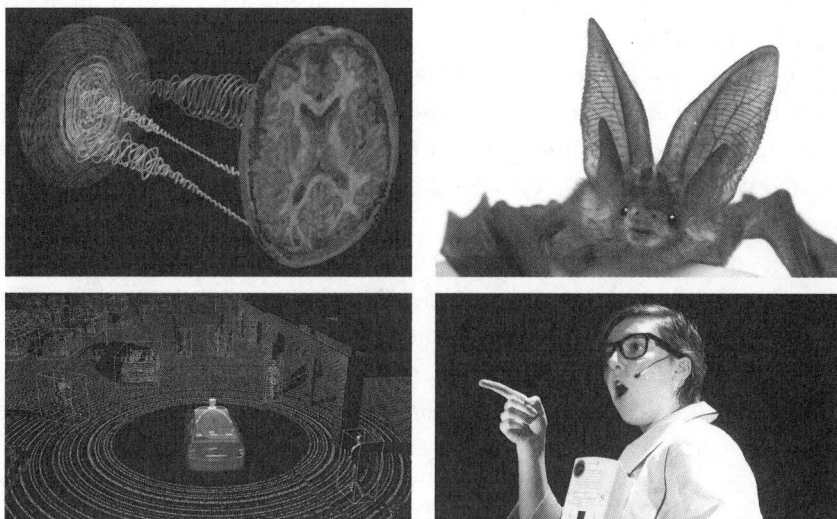

**FIGURE 9.1** Creating sensory data streams in technology and nature. (*Top left*) Magnetic resonance fingerprinting. (*Top right*) A bat's dynamic ears, which help to make it an expert navigator. Bottom left: LIDAR mapping the surroundings of a self-driving car. (*Bottom right*) My daughter, Hannah, engaging the audience's senses, and her own, in a production of the musical *Rent*.

*Sources*: (*Top left*) Courtesy of Dan Ma and Mark Griswold; (*top right*) Eric Isselee/Shutterstock; (*bottom left*) courtesy of Graham Murdoch (https://mmdi.co.uk/); (*bottom right*) Random Farms Kids' Theater, 2019.

construct, we find the seeds of something else entirely: something busy and surging, seemingly helter-skelter but also carefully coordinated. This is life as we know it. We might as well image that way.

☙

OK, so modern tomography is beginning to open up to more diverse ways of gathering projections, but are such approaches truly new, or are they just rearranging deck chairs, as it were? The history of tomography is full of clever new arrangements. I would argue that, at a deeper level, the new techniques I've surveyed here constitute new approaches to time in imaging. So far in our story of imaging, we have largely focused on spatial localization: how to tell here from there. In the process, we have not really accounted for the way most things move while we are busy sorting out where they are. In other words, we have given short shrift to time. As we know from Einstein, space and time are sometimes difficult to separate.

"Come on now," a critic might legitimately object. "This talk of space and time is all very philosophical, and maybe it is useful for slow imaging techniques like tomography, but is it really relevant to the way we actually see? Don't we just fix our eyes on the horizon and take in whatever vista presents itself to us?"

As it turns out, no, we don't. I recently attended a lecture titled "Seeing Space in Time" by a noted neuroscientist who specializes in vision. The speaker, Professor Michele Rucci of the University of Rochester, argued that visual scientists have historically spent too much effort on space and not enough on time. Many studies of vision have explored how the brain processes static images presented to the eyes. As Rucci and others have shown, however, no image is actually static. In fact, our eyes are always moving. You generally won't notice this in yourself. If you look closely at other people's eyes, though, you will see that they jitter and drift. At intervals, they also execute discrete jumps called saccades. Such eye motions are not exceptional—they are very much the norm. According to recent research, the average person racks up more saccades than heartbeats in a lifetime.

As an imaging device designer, I would be sorely tempted to view such motions as obstacles to be overcome. I might come up with stabilization techniques and motion compensation strategies to make sure the resulting images weren't too fuzzy. But evolution appears to have been more open-minded. It is now believed that eye motions are a deliberate approach used by the brain to enrich its store of visual information. Our eyes actually appear to encode information in both space and time. Imagine you're at a zoo (or, if you're a little more adventurous, on an open savanna) and you have a zebra in your sights. The contrast between black and white stripes will leap out at you, particularly as the zebra moves across your field of view. Now imagine that your eyes are jittering back and forth across the black-and-white pattern of the zebra's flank. Even if the zebra is standing stock still, the pattern on your retinas will be shifting from black to white and back again, over and over in time. From your brain's point of view, this is an even more distinctive signal. It is probably a signal you could pick out even if the zebra were otherwise well camouflaged. As long as your brain can keep track of where your eyes are pointing, targeted eye movements actually increase your visual acuity.[8]

Our jittery eyes are just one example of how nature uses time as a friend rather than as a foe. Our sense of smell is also known to take advantage of how odors enter our nose over time. Hearing, which sorts out oscillations of air pressure or vibrations of other materials at various frequencies, is all about patterns in time. In fact, encoding information in both space and time is a hallmark of all our senses. Our eyes offer us continually shifting views, like the continually shifting projections in GRASP. Our noses sniff out distinct aromas by identifying time-varying molecular signatures, like the complex signal patterns in MR fingerprinting.

These similarities lead naturally to the question of how else we can emulate our shifty eyes when it comes to human-made imaging. In radio astronomy, the earth's motion is already used to gather additional projections that would not be available if radio dishes were standing still. In the realm of optical imaging, significant recent attention has been paid to the development of so-called active cameras. Such cameras don't just receive light passively and convert it into images. Rather, they project their own patterns of light onto objects in their field of view and record how the patterns distort when bouncing off those objects. Like the projections used in medical imaging, this kind of projection can provide information about the three-dimensional shape of objects, since different shapes will distort structured light patterns in different ways. Many modern 3-D cameras go one step further, including a form of echolocation like what is used in ultrasound. Light beams are projected onto a scene, and sensitive detectors record minuscule differences in the return time of reflected light, which carries information about the distance traveled. The Kinect cameras in gaming consoles work this way, as do the LIDAR cameras installed in self-driving cars (figure 9.1, bottom left). Rather than just sweeping light beams repetitively back and forth like a classic radar range finder, though, modern active cameras vary their sweeping patterns over time in an approach called coded illumination. Coded illumination has been shown to increase visual information content as compared with static illumination or simple sweeps. It is the artificial equivalent of our eyes' saccades.

Unless you hail from the planet Krypton, you are probably not capable of emitting probing beams of illumination from your eyes. Nevertheless, your humble human eyes have been encoding information in space and time since you opened them for the very first time. Even if you were born without a traditional visual system, your other senses have been mapping out the world in remarkable detail from a variety of perspectives. In fact, diversity of perspective is one more lesson we can take from nature. In addition to our two eyes, our bodies come equipped with a pair of ears; a forest of branching, tactile nerve endings; and a dense tangle of taste buds and olfactory receptors. Imaging machines have only barely begun to follow suit. I'm not just talking about machines that use arrays of detectors to record projections as our retinas record light rays. I'm talking about machines equipped with ancillary sensors like electrocardiograms and pulse oximeters and cameras to track patient motion. From one perspective, combined PET–CT and PET–MRI scanners are multisensory devices. Nothing to date begins to approach the diversity of sensors in our own bodies, though. If emulating the senses is the name of the game, we might really want to give some thought to emulating our multifaceted sensory nervous system.

Here, then, is a glimpse of what imaging in the future may entail. The future I see is not one of regimented projections marching along to an autocratic beat. Instead, I envision a blossoming of continuous, coordinated, multisensor, multisensory imaging. I see a future in which we avoid repeating ourselves, using time to our advantage rather than trying to race it to a standstill. Our approach to imaging may still be smooth and methodical in some circumstances, but in others it may be jittery and jumpy, depending on what we wish to accomplish. Like the imaging methods nature has devised, our artificial imaging will consist more and more of using complementary sources to build up a dynamic picture of the world in space and time.

In drawing lessons from biological senses, I have focused on the sensory mechanisms we humans possess, but human senses are far from the most acute senses the world has seen. Perhaps we should cast the net wider than our own species.

Consider the bat (figure 9.1, top right).

If you want to be amazed, take a little time to read about how bats hunt.[9] You know, of course, that they use echolocation, which is a model for the human technologies of sonar, radar, ultrasound, and LIDAR. They sure don't seem to use it the way we do, though. They do not appear to form images of their environments laid out neatly like ultrasound displays. Instead, they create a complex and dynamic soundscape from which they pick out distinctive signatures, like you might pick out individual instruments in a symphony. Bats adapt both their vocalizations and their listening strategies to maximize useful information content. They adjust the frequency, intensity, and directionality of their chirps depending on their environment. Unlike many humans, they are also active listeners. It has recently been discovered that some bats move their ears back and forth like hummingbirds' wings. This motion is fast enough to generate significant Doppler shifts (i.e., velocity-dependent changes in tone like what you hear when a train rushes past you), and the corresponding tone signatures are believed to help bats localize their prey. Move over, shifty-eyed humans: bats, with their changeable chirps and fluttering ears, are virtuosos when it comes to using time in imaging.

Most impressively, bats do all of this on the fly—literally. They home in on the distinctive drone of a tasty moth, and they dive for it at top speed, focusing and modulating their own tones the closer they get. All the while, they somehow manage to avoid being shredded by thorny underbrush or dashing their brains out on a tree trunk. This virtuosic display of self-navigation puts the most futuristic self-driving cars to shame.

How do bats accomplish these feats? How do they generate, detect, and process a continuously changing stream of sensory data? They use a not-so-secret evolutionary weapon. Behind their protuberant ears, the comparatively small space of their skulls is packed tight with processing power in the form of their brains.

We too navigate our world with a heavy reliance on our brains. This is the only way we can make sense of sensation. When we take a walk through city streets, we are assailed by sensory signatures: street signs, traffic lights, automotive and human traffic—all the familiar sights and sounds and smells of metropolitan life. We hike through woods to the accompaniment of alternative choruses of sensory messages. In the dim recesses of a concert hall, a symphony of information washes over us. Beckoning marquee and bustling stage both call out to our senses, making them sing and dance (figure 9.1, bottom right). No matter where we go, no matter the time of day, space and time are tangled up in a dense web of mixed signals—what Ed Yong calls "the coursing chaos of the world."[10] We use our brains to sort out what all this sound and fury signifies.

While subjects posing for snapshots were once enjoined to "act natural," their careful poses never really reflected what we see in nature, or how we see it. Instead, the sensory mechanisms we first evolved as sea creatures have all along been built to bring us "staccato signals of constant information." We have also evolved powerful ways to make sense of the resulting information stream. It is time, then, for our imaging devices to take another cue from our biology. It is time for us to stop merely imitating our eyes and start emulating our brains.

# 10

---

# Emulating the Brain

## *Artificial Intelligence and the*
## *Future of Imaging*

How AI will not just interpret our images but change them.

T he world we see is a lie.

Back in chapter 4, I tried to convince you that tomography—the art of reconstructing cross-sectional images from projections—is a direct analogue of depth perception in our own natural vision. I never really told you how many projections are required to get a good image, though. In fact, a rich body of theory addresses this question precisely, and various mathematical and practical criteria for sufficiency have been worked out over the last century or so.

Human vision takes all this nice theory and throws it completely out the window. Generally speaking, humans have two eyes. This means that we get two separate views to work with in constructing a model of the outside world. Each of these two views is a projection of all illuminated objects within our field of view onto the surface of one of our two retinas. In other words, we are working with only a single pair of two-dimensional projections to reconstruct a full three-dimensional space.

From the point of view of image reconstruction, two projections are not nearly enough. Imagine trying to generate a cross-sectional image by back-projecting just two projections.[1] Now imagine that your two projections are taken from nearly the same angle (as is the case when one is looking to the horizon from eyes separated by only a few inches). With such similar projections, you should barely be able to nail down the location of anything at all.

Let's make matters even worse. If you can, try a quick experiment for me. Close one eye and look around. You still have a pretty good sense of how far away everything is from you, don't you? You can tell what is within arm's reach and what is off in the distance. If you think about it, that is patently absurd. With only one eye open, you are working from just one projection. You should have no information whatsoever about depth along your line of sight. And yet, apart from some minor sense of disorientation perhaps, your picture of the world in all its three-dimensional glory is intact.

That picture of the world, which seems so real to you, is an optical illusion. Philosophers have long questioned the nature of our perceived realities, and now the mathematicians have reason to join them. People who struggle with disorders of depth perception know this truth from practical experience. The visible world we appear to inhabit is a construct. It is an educated guess.

To be fair, if visual reality is a guess, it is rather a good one. Our eyes, and our estimates of distance, do a pretty good job of keeping us out of the way of onrushing vehicles and safely removed from the edges of cliffs. How, then, do we do it? How do we guess so well? That is where things get really interesting from an imager's point of view. The built-in image reconstruction engines in our heads are continually bringing all kinds of additional information to bear. For example, we use not only space but also time to our advantage. As discussed in chapter 9, our eyes are always roving even though our brains preserve the illusion of stability. This gives us additional information about how things change over time. We can also move our heads, quietly gathering a variety of additional projections without really thinking about it. This is one of the benefits of working with sensory data streams.

We also rely on diverse sources of information other than just the raw evidence of our eyes. By the time we are able to walk, we have assimilated a rich body of knowledge about how the world works. We know how big people are compared to everyday objects like pebbles or automobiles. We know that if that car over there looks a lot smaller than this person over here, then the car is probably farther away from us than the person is. We also understand perspective. If a road appears to taper down to a point, we know that it is probably not actually getting significantly narrower but is instead extending off into the distance. In fact, we automatically adjust for perspective, identifying common points of reference in multiple views of

familiar objects. We use other subtle cues about three-dimensional structure as well, like the way light reflects off objects, and the way objects in
the foreground of our vision tend to block objects behind them. In other
words, we depend heavily on models of the world that we have learned
over time. Whether or not we have studied physics in school, we have an
instinctual grasp of the laws of physics, at least when it comes to the behavior of ordinary objects. We understand the normal disposition of things in
space and time, and we use this knowledge to invest our woefully insuf-
ficient two-dimensional projections with realistic depth.

This means that our view of the world is based heavily on assumptions.
Assumptions are what allow us to appreciate the grand sweep of a sublime
mountaintop vista. Assumptions are also what keep us from bumping into
things all over the place. The fact that we dare to venture out into the world
at all with such a thoroughly artificial sense of where things are is actually
quite remarkable. Despite our paltry pair of projections, we forge ahead,
damning the torpedoes, keeping our foot on the gas because we know that
we will not reach that distant stoplight for some time yet.[2] When we were
newborns, we lurched and tottered because we did not know how to judge
depth. Now we swagger onward because we have convinced ourselves that
we are never out of our depth.

Most of the time, our lived experience serves us very well indeed. Since
we rely so heavily on prior knowledge, though, we can be relatively easy
to fool visually. Witness the long list of clever optical illusions we love to
gawk at. Delightful compilations of mind- and vision-bending optical illusions can be found online and in print. In various cities around the world,
there are entire museums devoted to such illusions. Figure 10.1 shows an
example from the Museum of Illusions in Manhattan. The tableau in the
photo is specially designed to fool our natural depth perception. My son,
Noah, who is standing, appears to tower over his older cousin Zoe, who is
unnaturally dwarfed by the chair on which she sits. In actuality, the chair
is unusually large and is located far back in the room. Noah and the two
posts that appear to be the front legs of the chair are in the foreground, and
Noah's left arm is actually resting in air, the chair's back leg far behind him.
This artificial scene takes advantage of our brains' assumptions about size
and distance to create an amusingly false impression.

Speaking of impressions, we have yet to address the most successful
engineered optical illusion of all: art. Paintings and their lifelike cousins,
photographs, allow artful humans to represent three-dimensional scenes

**FIGURE 10.1** Fooling the brain. A careful construction in New York City's Museum of Illusions uses deceptive cues for depth perception to create an unnatural impression of size.

*Source*: Daniel K. Sodickson.

in resplendent detail. Yet paintings and photographs are flat. It is our brains that assign depth to them. We are not completely fooled, of course: we don't generally try to walk into painted scenes. We are typically attentive enough to recognize subtle signs of artifice (like an ornamental frame surrounding the scene), or else our binocular vision gives us dynamic cues, as we approach, that the perspective in front of us is false.[3]

If you work at it, you can sometimes force yourself to make life resemble art. Focus on a real-life scene in front of you with sufficient determination, and you may be able to see it as a painting, ignoring everything you know about relative sizes and perspectives and flattening out your perception of depth. It's disconcerting.

Why have I been working so hard to deconstruct your view of the world? My goal has been to highlight an essential fact about natural vision. The raw data from our eyes is only part of the picture. The rest comes from our brains.

What can we learn from this? If we really want to bring our artificially augmented vision to the next level, then perhaps we should aim not only to imitate our eyes but also to emulate our brains. Fortunately, artificial brains are not just the stuff of science fiction nowadays. They are rapidly making their way into real science, in the form of artificial intelligence.

AI has already made a rather dramatic entry into the field of radiology. The following words, uttered during a 2016 conference on machine learning and the market for intelligence, sent shock waves through the field: "Let me start by just saying a few things that seem obvious. I think if you work as a radiologist, you're like the coyote that's already over the edge of the cliff but hasn't yet looked down so he doesn't realize there's no ground underneath him. People should stop training radiologists now. It's just completely obvious that within five years deep learning is going to do better than radiologists because it's going to be able to get a lot more experience."

Just to be clear, this is *not* what I had in mind when I said that we should aim to emulate our brains. Nevertheless, the speaker, Geoffrey Hinton, has an unquestionable track record when it comes to intelligence, both natural and artificial. In 2016, he was already recognized as one of the key figures behind the modern AI boom. In 2018, he would go on to win a Turing Prize, which is the closest thing to a Nobel Prize offered in computer science. In 2024, he would collect a Nobel Prize in Physics as well. Back in 2016, however, his offhand pronouncement made him medical imaging's greatest boogeyman.

Vigorous objections to Hinton's prediction surfaced quickly. Working groups were convened and pointed position papers published. For a couple of years, it seemed that at least one session at every radiology-related conference I attended was devoted to assuring anxious radiologists that they had nothing to fear but fear itself.

Over time, as applications of AI in medical imaging became increasingly common—without, as of yet, eliminating any jobs—a calmer and more collaborative picture of the future developed. AI algorithms would be deployed to make human radiologists better at their jobs, and also more satisfied with them. Neural networks would be pressed into service as additional sets of eyes, catching errors and boosting confidence. They would be deployed for triage, handling without fanfare the reams of normal scans that currently fill much of radiologists' time, and flagging only the worrisome findings for review by human experts. Intelligent tools would help to reduce the sheer number of image pixels that human radiologists were called upon to inspect, allowing radiologists to go about their business less mechanically and more humanely. In short, humans and machines would work together to improve patient care while at the same time helping to humanize the field of radiology. My colleague Curt Langlotz at the Stanford Department of Radiology had a pithy summary for this new predicted symbiosis: "AI will not replace radiologists. Radiologists who use AI will replace those who don't."[4]

The noted physician and medical futurist Eric Topol devotes an entire chapter of his 2019 book *Deep Medicine* to describing how AI can liberate doctors who deal with patterns.[5] Radiologists, pathologists, and dermatologists are all experts in the extraction of medical information from images: radiologists from scans, pathologists from microscope slides, and dermatologists from photographic representations of skin. Topol argues that the dehumanization of such specialists began long ago, when the pressures of modern medicine pulled them away from medicine's deep roots in interpersonal connection. He suggests that if radiologists in particular are willing to yield up some of their accustomed tasks to AI algorithms, it will free them up to reengage with patients in a way they haven't done since radiographers took over the everyday duties of scanning. He believes that in the future, radiologists will be able to emerge from the salt mines of their subterranean reading rooms to engage anew with other doctors. He imagines that today's masters of the arcane imaging arts can become tomorrow's "master explainers," making complex findings understandable for patients and doctors alike.[6] This is just one way in which, as Topol's subtitle frames it, "artificial intelligence can make healthcare human again."

Returning to Hinton, the vigor of the debate that erupted following his inflammatory words reflects the emerging power of AI, which is poised to disrupt many fields of human endeavor and, in the process, either replace or redefine the roles of any number of skilled laborers. At the same time, both sides of the debate illustrate powerfully the perils of prediction.

On the one hand, the state of affairs Hinton predicted has clearly failed to come to pass. As of the time of this writing, it has been nine years since Hinton made his prediction, and deep learning isn't anywhere close to replacing radiologists.[7] In fact, I am not aware of a single radiologist whose job has demonstrably been taken over by machines. Though a mad rush of companies and medical institutions are busy developing algorithms for automated image interpretation, disease detection, and risk prediction, only a comparatively small number of AI algorithms have been approved for clinical use, and they have barely budged the day-to-day paradigm of medical image interpretation. This is partly a reflection of the measured rate of adoption of new technologies in medicine. Any new medical technology is subject to careful protective regulation and, even when approved, it must survive the stress test of busy clinicians' everyday workflow.

The failure of Hinton's prediction, though, can also be traced to a common misconception about what radiologists do. They don't just hunt mechanically for a tumor and sound the alarm when they find one. Neither do they simply sift images for a definitive diagnosis. The radiologists I have observed—and I have observed many over the years—commonly employ a powerful mix of detective work, free association, and informed judgment. Whereas computers are still grossly inept at the children's game of guessing "what's wrong with this picture," human radiologists are its grand masters, their eyes going right to what is out of place even if they have little in the way of clinical suspicion to go on.[8] They flip from one image type or imaging study to another, often faster than even an image-savvy observer like me can follow. They follow suggestive leads hither and yon, adjusting their estimation of potential diagnoses on the fly. They are skilled at clarifying what they don't know as well as what they do, and they call for additional tests when they need more clarity. Moreover, unlike artificial neural networks, which may need to digest thousands or millions of examples before learning to recognize any particular image feature, a trained radiologist will often remember a new imaging finding forever after seeing it just once. In addition to all their specialized training, meanwhile, radiologists also have access to the vast pool of human intuition known as common sense. AI may one day master such skills. Not today, though, or even tomorrow. According to another Turing Prize winner, Yann LeCun (a colleague of mine at NYU and Meta's chief AI scientist), "the smartest AI systems today have less common sense than a house cat."[9]

From this perspective, a lot of the AI buzz in radiology today amounts to herding cats. The lion's share of today's AI applications in medical imaging involve relatively straightforward tasks of image interpretation: detecting suspicious lesions in x-rays or CT scans; automatically identifying and delineating anatomical structures in CT, MRI, PET, or ultrasound; making some data-driven predictions of adverse outcomes (say, who will require intensive care for COVID-19 in a few days' time or who might develop cancer in a year or two). Applications like these can be helpful in guiding care decisions, but they cannot yet substitute for human judgment. (I will address the ascendency of large language models and foundation models, which purport to emulate judgment, later in this chapter.) Most of these applications involve some kind of pattern matching, and most are accomplished through the brute-force processing of large datasets. Meanwhile, many applications of AI in imaging continue to be hounded by concerns about generalizability, explainability, and bias implicit in the selection of training data.

Even if all of these challenges can be overcome (as in all likelihood they can, perhaps even relatively soon), visions of a future in which machines merely compete or collaborate with human radiologists in their accustomed tasks misses the point, I think. Such a prediction grossly underestimates the potential power of AI to transform medical imaging. Because AI can do much more than merely copy well-defined human actions like identifying features in images. It can also emulate some of the deeper functions of our evolved biology. These functions include the magic of depth perception with which we began this chapter. They also include powerful capabilities like memory and predictive power and the coordinated processing of sensory data streams. These are the built-in capabilities that allow us to navigate our world as if we know what we are doing.

In the remainder of this chapter, I will survey just a few types of powerful new transformations that AI can bring to the table, laying the groundwork for new imaging technologies and new uses of image-derived information.

Artificial intelligence can be defined broadly as any use of machines to mimic naturally evolved intelligence. This definition casts a pretty wide net, since humanity has devised many kinds of mechanical constructs, and since *intelligence* itself is a catch-all for many disparate features and behaviors that one might seek to emulate. The use of computers to emulate natural intelligence dates back to the beginning of the computer age, and this kind of AI has undergone numerous stages of ebb and flow over the last century. We currently inhabit a time of undeniable flow, with AI cropping up just about everywhere in commerce, science, and public discourse. This is in large part due to a particular branch of AI called machine learning and a particular branch of machine learning called deep learning. Machine learning encompasses a wide array of techniques that allow computers to improve their performance with experience. Rather than being hard-coded or hand-tuned in one fell swoop by humans, machine-learning algorithms adapt. Deep learning, meanwhile, is so called because it makes use of powerful computing structures called deep neural networks.

As I mentioned briefly in chapter 8, artificial neural networks in general are modeled loosely on biological networks of brain cells. The neurons in our brains, and in the brains of many terrestrial organisms, tend to be organized in connected groupings or layers. Information is passed from one layer to the next, with the output of various neurons in one layer combining to form the input to neurons in the next layer. Over the years, it has been shown that many types of computation can be broken down into successive stages that fit this layered model. Deep neural networks tend to use many internal layers, which increases their complexity but also their capacity for learning.

The modern era of deep learning had its start in the 1990s and built up steam in the early 2000s. Interestingly, imaging has played a key role in deep learning from the beginning. Some of the earliest deep neural networks were trained to recognize low-resolution images of handwritten digits. A notable breakthrough in the field came in the early 2010s when deep neural networks were trained to identify the content of images from a large public dataset called ImageNet. Computer vision remains an essential driver of advances in deep learning, as well as a significant beneficiary of those advances.

For those interested in digging deeper into deep learning, many wonderful online and print resources exist nowadays, at whatever level of detail you prefer. If you are scientifically inclined and would like to understand the fundamentals of deep learning methods, I heartily recommend a now-classic review titled simply "Deep Learning," published in *Nature* in 2015.[10] (This review was coauthored by the three thought leaders who went on to win the Turing Prize in 2018: Yann LeCun, Yoshua Bengio, and Geoffrey Hinton.) Our concern here, though, is the future of seeing. I will therefore provide you with a simplified summary of some key aspects of deep learning that could be truly transformational for imaging.

Figure 10.2 illustrates the basic structure and function of one popular type of deep neural network called a convolutional neural network, or ConvNet. ConvNets have been at the heart of the deep learning revolution,

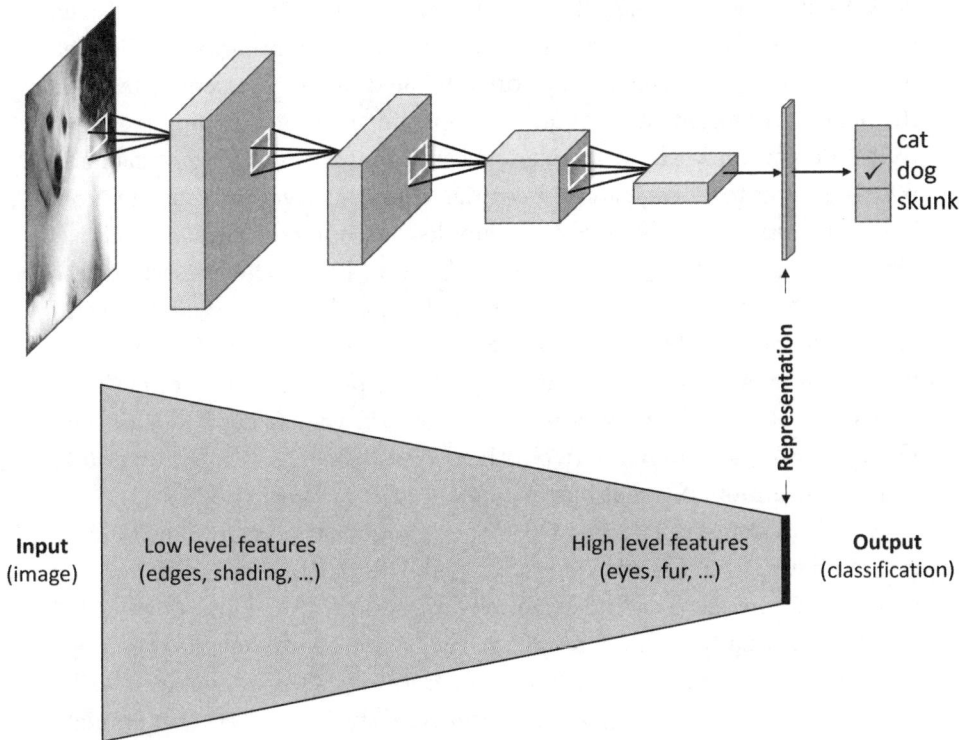

**FIGURE 10.2** Deep learning 101: a convolutional neural network for image classification.

*Source*: Created by the author, incorporating an image of an American Eskimo Dog (en:User:StarSaber, CC BY-SA 3.0, via Wikimedia Commons).

particularly for computer vision applications. They also bring us back to Yann LeCun, who is credited with inventing them around 1990. I've spoken often in this book about the value of drawing inspiration from biology. Well, according to Yann, ConvNets are modeled directly on the way vision is managed in biological brains.

It is probably no coincidence that Yann has compared AI with house cats, because we actually owe a lot of what we know about the neuroscience of vision to cats. Pioneering work in the 1960s showed that there is a hierarchical structure to the way cats see. Visual information passes from layer to layer in their brains in a pattern of successive refinement that is now known to be present as well in the brains of humans and other primates. In early layers, closer to the eye, adjacent pixels from the raw projections captured by the retina are compared. Edges are highlighted, and orientations are sorted out. These primitive features are later concatenated and combined into higher-level features spanning more of the full field of view. In other words, the parts of the brain devoted to vision are organized for a process of successive distillation. Raw visual data points are distilled down to highlights, which are in turn distilled down to essential features that a creature might care about, like appealing eyes or alarming teeth.

Even creatures with far simpler eyes and brains than those of cats and people are able to detect features essential to their survival and reproduction. Take the *Limulus polyphemus*, for example—the horseshoe crab, which has existed in nearly its current form for millions of years. The horseshoe crab is a close relative of the now-extinct trilobite, and it has served as a valuable nonfeline model for studying the neuroscience of vision. The primitive layered visual system attached to the crab's compound eyes is remarkably well adapted to identifying other crabs, even in watery depths or dark of night. Hierarchical neural networks have clearly been worth holding on to from an evolutionary point of view.

As shown in figure 10.2, the basic operation of a ConvNet follows a process similar to what happens in crab, cat, and human brains. Input data—in this case, an image of a fluffy Samoyed or American Eskimo dog—is fed into the starting layer. In subsequent layers, adjacent pixels in the image are compared using simple filtering operations called convolutions (the origin of the "Conv" in "ConvNet"). Other thresholding filters that emulate the behavior of biological neurons are also applied. Pixels in the filtered images are then pooled together and collapsed into more compact representations. The process continues with successive stages of distillation.

The end result is a compact set of high-level features, which are then used to guess which of a set of possible labels best applies—say, *cat, dog,* or *skunk* in this schematic example. The overall effect is of an information funnel, as shown in gray at the bottom of the figure. The input image is progressively distilled into ever-higher-level features that eventually result in an output classification.

Where is the learning in all of this? Learning comes when you iterate the process with many different images whose correct label is known. In each case, you compare the network's guess with the correct answer. If the guess is far off the mark, you move backward through the network and adjust the strength of connections between layers using a process called backpropagation. You can think of backpropagation as a kind of analogue to back projection in tomography.[11] Back projection spreads each new projection back across the image to refine your estimate of internal structures. Likewise, backpropagation spreads each bit of new information back through the network, refining your ability to estimate labels correctly. Just as in tomography, the more times you perform the process, the better your estimates get. Your network learns.

While the underlying components of neural networks like ConvNets are relatively simple—generally quite a bit simpler than biological counterparts like neurons—they have remarkable flexibility when they are brought together en masse, and when they are fed with large quantities of training data. As a result, uses of ConvNets have positively exploded in recent years. They have been used for automatic labeling of photos online or in your smartphone collection. They have been trained to identify objects in real-time video streams, such as those used in self-driving cars. They have also been used to classify many kinds of input data other than images.

Many types of artificial neural networks other than ConvNets have also been developed.[12] There are multilayer perceptrons, which hark back to computing structures explored in the mid-twentieth century. There are recurrent neural networks, which emulate some aspects of biological memory. Some recent hot commodities are called transformers (not to be confused with the shape-shifting alien intelligences of the same name featured in a popular movie franchise). Transformers are particularly good at sorting out relationships in time and space, and they have had a notable impact in areas like natural language processing and speech recognition, as well as computer vision. Chains of interconnected neural networks are sometimes trained together or in direct competition with one another.

One noteworthy competitive pairing is called a generative adversarial network (GAN), which has proven to be particularly adept at generating "deep fakes." GANs can create plausible-looking images that convincingly distort true scenes or that appear out of thin air with no immediate connection to the real world at all. In this realm of so-called generative AI, GANs have recently been outpaced by diffusion models, which distort input data in a series of small steps.

Neural networks also have an ever-widening range of applications beyond computer vision. For example, they are used routinely to identify trends in weather or business, to understand and produce natural language, to play games, to design new drugs and materials, and even to discover new mathematical algorithms. The pace of development right now is nothing short of pell-mell, and an ever-expanding community of computer scientists and engineers is jumping into the breach. Papers posted online are often cited many times before even making it through peer review to be published in final form. Companies large and small are scrambling over one another to bring AI-driven products to market. This collective churn has all the historical hallmarks of life on an expanding frontier. The air is positively humming with possibility, and with peril.

In due time, we'll explore at least some of what makes the AI gold rush perilous. For now, though, let's consider the possibilities from the vantage point of imaging. What neural networks offer, in a nutshell, is a new kind of transformation. As you have seen, new transformations can be incredibly powerful. The history of imaging has arguably been driven by a progressively more complex set of transformations. Now we have something still more complex, and potentially still more powerful, at our disposal. We can take raw input data, operate on it with a series of layered operations, and end up with highly refined outputs that may look nothing at all like the input but are nevertheless full of useful information.

So, how can we use these new transformations to rethink how we image?

One possibility that leaps immediately to mind is to use neural networks to search images for specialized information beyond what kind of cute pet they may contain. We can comb medical images for subtle signs of worrisome diseases, like cancer or multiple sclerosis or stroke. We can inspect microscope slides to find corresponding indications of cellular pathology.

We can process large volumes of astronomical data to find promising candidates for planets that may be circling around stars other than our sun. All these areas are currently being pursued with vigor. Automatic pattern recognition may not yet eliminate the need for human eyes and brains—let's learn from Hinton's example and not be too quick to cry wolf (or Wile E. Coyote) when it comes to radiology or microscopy or astronomy. However, pattern recognition algorithms can certainly help us sort through far more image data than we once could. Neural networks can help draw our eyes to patterns that might otherwise be difficult to see. They can even help us to discover brand-new patterns.

And yet, given everything I've just told you about the deep learning revolution, doesn't setting our sights only on recognizing patterns in images seem just a little . . . uninspired?

I do not mean to cast any aspersions whatsoever on current AI-based approaches to automated pattern recognition. This is a rich field of study, with deep theoretical foundations and many applications of genuine practical value. That said, such approaches tend to treat input data as given, as they focus on finding new ways of digesting that data. If you have learned anything from our journey together so far, though, it should be that imaging data is not simply given. We can gather it in many different ways, depending upon our reasons for generating images in the first place. So, how can AI help us to gather different imaging data? How can it help us to image differently?

Such questions naturally push us further upstream in the imaging process. We already know that, in addition to helping us deal with large volumes of imaging data, AI can help us to do more with less. Recall from chapter 8 that AI is currently being used to reconstruct images from limited projections. Such approaches help to increase shutter speed in settings where speed is of the essence. It has even been shown that neural networks can learn which projections are most important to gather if you want to get the best image quality from a limited number of projections.

It is also well documented by now that AI can help to rescue images that have been gathered under various less-than-optimal conditions. Many smartphone cameras nowadays can automatically brighten up poorly lit photos. Recent camera models have been trained to portray darker skin tones accurately, addressing long-standing inequities in the digital representation of people around the world. Neural networks trained to remove noise or undesired artifacts from images are beginning to be used in many

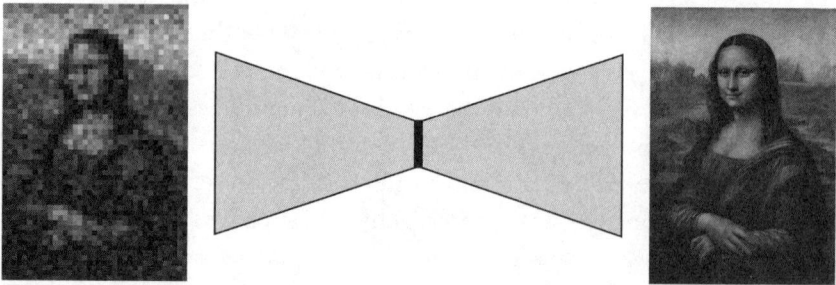

**FIGURE 10.3** A neural network for image enhancement.

*Sources*: (*Left*) Generated by the author from the image on the right using MATLAB (The Mathworks Inc.); (*right*) Dianelos, CC BY-SA 4.0, via Wikimedia Commons.

different fields of imaging. Once again, there is a clear biological precedent: our brains do this sort of thing all the time. We can see in low-light conditions. We can piece together scenes from behind partial obstructions like tree branches or porch screens.

What kind of neural network would you use if you wanted not merely to label an image but to modify it—to fill in missing information, to highlight particular regions, perhaps to sharpen things up all around? Figure 10.3 shows a sample network arrangement that can perform tasks like these. Such a network is called an autoencoder. It begins with an information funnel similar to what is used in a ConvNet. In an autoencoder network, this first stage is called the encoder. The encoder is followed by an inverted funnel, not surprisingly called a decoder, whose job is to reverse the process of distillation and regenerate an image. The encoder distills an image down to its essence. The decoder then produces a modified image from that essential representation. Autoencoder networks can be trained to perform many types of modifications. They can remove some of the nonessential "fuzz" of noise, as shown for the *Mona Lisa* image in the figure. They can increase image resolution, as is also illustrated in the figure. They can fill in gaps in images, a task known rather evocatively as inpainting. They can change the overall appearance of an image in a process called style transfer, which can be used to render the *Mona Lisa* as, say, Van Gogh might have painted her. Not only style but also image quality can be transferred. For example, neural networks can be trained to take cheap, low-quality images and make them look more like images obtained using top-of-the-line equipment. All you need is a training dataset of low-quality images paired with corresponding high-quality counterparts. Denoising, resolution enhancement, and image quality transfer have already helped to enable the less-is-more trend in medical imaging. Not satisfied with the image quality of your small, low-field MRI system? Then tune it up with neural networks.

Let's not get too carried away just yet, though. It is important to remember that AI-based image enhancement approaches are not without their limits. In fact, they hit some limits rather quickly. If you push a denoising autoencoder too hard, it will start to blur out important image details. If you rely too heavily on image quality transfer, you risk ignoring unique properties of the thing you want to see in favor of your assumptions about how an image of that thing should look.

The example of depth perception shows us that our brains routinely make all kinds of assumptions about how things should look. In artificial imaging, however, we tend to be rather cautious about introducing too many potentially unfounded assumptions. Would you really want your self-driving car to plow into an intersection based only on its estimate of the relative sizes of things?[13] Astronomers are justifiably leery of seeing distant structures that aren't really there. Likewise for microscopists and the tiny structures they study. In medical imaging, optical illusions can do real harm. A key goal of medical imaging, after all, is to be able to locate *unexpected* structures in a known field of view. If radiologists or their artificial surrogates miss those anomalies, or hallucinate them when they are not there, it is patients who will pay the price. We can try to keep the creative tendencies of neural networks under control using various reality checks, as described in chapter 8. We can input all the relevant physics and mathematics that we know. Still, when our goal is to find every little anomaly in an image, even ones that might escape casual inspection, we really don't want to go as far as our brains do with depth perception.

It is when we shift our goals that we can begin to lean harder on machine learning. Sometimes we are not interested in pure, unbiased visual exploration. Sometimes we just want answers. Is there something in the sky that hasn't been there before? Has the tumor in my prostate gotten bigger? Does my mother have breast cancer? In such cases, it is already known that one can make do with less exhaustive imaging studies than would otherwise be required—rapid sky surveys in astronomy or abbreviated medical imaging protocols. In recent times, AI researchers led by my NYU colleague Sumit Chopra have found that in MRI, one can also get away with far fewer projections or far noisier projections than would normally be required to make a pretty image—as long as one is using a neural net to make a particular diagnosis.[14] My graduate student Radhika Tibrewala built a tool to simulate distorted images emerging from low-field, accessible MRI machines in order to demonstrate the same thing for other types of image quality degradations.[15] Machines, it would seem, can sort through lots of chaff if they have been trained to hunt for wheat. We may not enjoy looking at the messy images that result, but in this case we really don't need to look. We can skip right over the images and go directly from the raw data to the answer.

To be clear once again, I am not talking about eliminating images in general, or replacing the human consumers of those images. The need to inspect carefully constructed images for anything they might contain will not go away soon. In fact, I fully expect that we will forever be striving for images with ever greater clarity and information content. Instead, I am talking about using highly streamlined imaging data, together with the amped-up transformations provided by AI, to answer particular questions automatically. You may think of this as the equivalent of catching quick glimpses of surrounding traffic through a rearview mirror and counting on our brains to flag just what we need to see to keep us safe.

At the heart of many of the questions we might want to answer automatically with imaging is the question of change. Novas and supernovas are—literally—novel objects in the sky that were detected and studied by keen observers long before we had modern astronomical observatories. Modern motion detectors are deployed strategically to sense changes within their fixed field of view. Certainly, much of radiologic image interpretation involves a studied assessment of noteworthy changes. Just as we all carry in our heads a robust set of assumptions about how the world works, radiologists carry with them detailed knowledge of internal anatomy and its normal variants. While they are eminently capable of identifying many particular abnormal patterns, they also excel at spotting worrisome departures from normalcy, whether familiar or not. It has long been known that some of the earliest layers in biological neural networks associated with vision serve to highlight edges, the spatial transitions from one thing to another in our field of view.[16] We are by nature attuned to change. The deeper layers in a radiologist's brain, meanwhile, constitute a carefully trained difference engine. When a radiologist casts her eyes on a set of images, she can often glean at a glance what belongs there and what doesn't.

In order to appreciate that something has changed, one needs to know what characterizes that thing as itself. Such a philosophical grasp of identity may seem far from the capabilities of today's artificial neural networks. From another perspective, though, this kind of understanding is a particularly good fit for AI. What many neural networks do, at root, is learn compact representations of things. Both the ConvNet in figure 10.2 and the autoencoder in figure 10.3 distill their inputs down to a simple set of core features before going on to generate their respective outputs. Precisely which features each network learns to distill depends on the task for which it is trained. What would we get, however, if we set a network to the task of simply determining whether something has changed or not?

This brings us to the blossoming area of "self-supervised learning." We call machine learning "supervised" when we feed it with correctly labeled training data, or when we present it with ground-truth output for every input in the training set. By contrast, *self*-supervised learning systems do not need to know the right labels; they figure things out on their own. The example in figure 10.4 illustrates how this can be done. The so-called Siamese network structure in the figure comprises two encoders that are trained in tandem. One encoder takes an image as input and produces a distilled representation at the end of the funnel. The second encoder is fed with a modified copy of the original image—say, the iconic *Mona Lisa* without her familiar shoulder-length hair.[17] The outputs of the two encoders are then compared to determine whether or not they represent the same thing. The process is repeated for many different images, and for many different modifications of each image, and backpropagation is applied in the same way to both encoders, pulling representations of similar things together while pushing representations of different things apart. At the end of the training, the encoders have learned to recognize features that distinguish images from one another, ignoring all the types of incidental modifications included in the training set. This brand of self-supervised learning is called "contrastive learning." Another common flavor of self-supervision is "masked autoencoding," in which parts of images, documents, or another suitable inputs are removed, and autoencoder networks are trained to fill in what is missing using context.

You can generate robust training sets for self-supervised learning without knowing anything to speak of about the content of images. You just

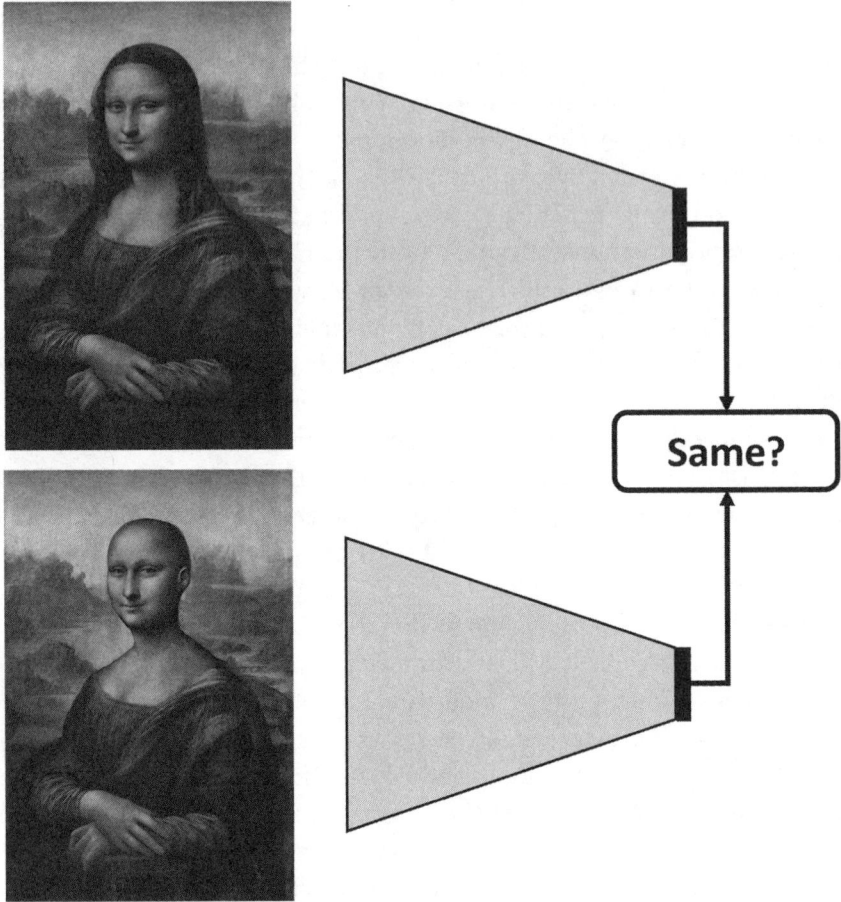

**FIGURE 10.4**  A neural network for self-supervised learning.

*Sources*: *(Top)* Dianelos, CC BY-SA 4.0, via Wikimedia Commons; *(bottom)* created by the author from the image above using the generative fill tool in Adobe Photoshop, with inspiration from a 2013 image used in a cancer awareness campaign by the Fondazione ANT Italia.

need to choose which modifications you want your network to learn to ignore, or to undo. To put it simply, self-supervised networks learn self-similarity. They learn what makes something what it is, ignoring irrelevant modifications. That general knowledge, though not directed toward particular tasks, has nevertheless proven to be a useful base for more targeted learning. Once an encoder has been trained using self-supervised learning, it can be removed from the Siamese network, the autoencoder network, or any other setting in which it was trained, and can be used as the first stage of other task-oriented networks, which then require only a little fine tuning to achieve state-of-the-art performance.

There is something deep and powerful lurking within self-supervised networks, though, beyond their general utility for multiple tasks. In making sure that representations of a whole remain intact even when some parts are modified or masked out, networks like the one in figure 10.4 get pretty good at identifying the essence of things. In the process, they also learn at least something about context: say, what is foreground and what is background in an image. They learn overall classes of things, too: animal versus vegetable versus mineral.[18] They can even learn what we might identify as general concepts. It is not too difficult to imagine, for example, that pairs of images like the ones in figure 10.4 could be a good starting point for learning the concept of "hair."

Self-supervised learning is a key feature of large language models like OpenAI's much-vaunted ChatGPT, which has caused such a stir since its initial release, and which has helped to spur a recent explosion in generative AI. ChatGPT and its cousins, which can produce volumes of complex content, have been trained on the comparatively simple task of predicting the next word in a text sequence based on words that have come before. In the process, they appear to have mastered some of the fundamentals of language. Indeed, large language models are sometimes referred to as foundation models. Such models, trained in a self-supervised fashion on gargantuan collections of data, are the foundation of many flavors of modern generative AI. In addition to pure-text foundation models, which can answer questions or write term papers on demand, there are models that have learned to associate text with images and can be prompted to generate realistic-seeming pictures or works of digital art. Both commercial and open-source tools now exist to generate convincing video clips that convey made-up messages using real people's faces and voices—whether authorized or not. Generative AI makes a virtue—and a vice—of hallucination. It's like a virtuosic and eager-to-please child, quick with a clever confabulation to delight the adults in the room.

I have had the pleasure of discussing self-supervised learning with Yann LeCun, whose enthusiasm for this red-hot area is at least as palpable in person as it is in his various lectures and writings on the subject.[19] He gets visibly energized when the topic comes up, as if electric current has begun coursing through his body. Why? The reasons are partly practical. One of the principal bottlenecks for supervised learning is the need to generate correct labels, but self-supervised learning doesn't need predefined labels. It can sort things out for itself, somewhat like an infant learning about the world. And this, I wager, is what excites Yann even more. Self-supervised learning appears to emulate how we humans actually learn. Yes, we have parents to tell us the names of things and teachers to pass on humanity's collective store of knowledge. Long before we learn how to process everything they have to teach us, though, we are busy learning from experience. We are constantly taking in information about the world around us and sorting out our place in it. We are digesting continuous streams of visual and other sensory data. We are learning object permanence and motor control and cause and effect. In other words, we are building up self-supervised models of how the world works.

When it comes to applications of self-supervised learning, Yann is not actually the greatest fan of generative AI, though his team at Meta has certainly produced powerful examples of the genre. He argues that the foundation models used in generative AI are not at root foundational, in that they do not glean much at all in the way of common sense about the world. They learn to generate content, he maintains, without genuinely understanding context. In this sense, they are shallower than we might imagine them to be. Yann allows that the output of generative AI systems is impressive, and in some narrow respects superhuman, but without supervision it can wander away from established facts. Whereas our intuitions are grounded in motivations like survival and the avoidance of pain, our computers have a substantially more fragile tether to truth. On the contrary, one might say that generative AI systems have been trained not so much to replicate as to exploit our own learned senses of plausibility. Generative networks master a childish form of mimicry without having a child's instinctive grasp of physical laws and interpersonal relationships. As a result, they cannot repurpose knowledge, or learn from limited data, or draw robust insights even from single examples as most human children can.

To address such limitations, Yann has publicly called for a shift from generative to objective-driven AI, something less performative and more goal oriented.[20] In this, he aims to emulate biology once again. In fact, he has proposed a provocative framework for building autonomous machine intelligence modeled loosely on human intelligence.[21] The idea is to emulate how we humans appear to operate as we navigate the world: formulating action plans driven by cost and reward and informed by sensory modules, which provide real-time feedback about where we are and how we are doing in achieving our goals. One essential component, he maintains, will be a central neural engine capable of juggling a vast collection of self-supervised representations and putting them to use for whatever the current objective might be. The core function of this engine is to collect and distribute models of the world.

It is world models like these, I believe, that provide the educated guesses that convert our eyes' meager 2-D projections into a full-depth picture of the world. Our hard-won instincts about physical law and rules of perspective do not remain in the ivory tower of our academic minds. They are brought to bear liberally in everything we do, and they underpin everything we see.

Let's consider, then, how we might use the world models emerging from self-supervised neural networks to change the way we do artificial imaging. At the very least, we should be able to use them to flesh out the projections provided to us by imaging machines, allowing us to do even more with even less. We should also be able to use some of this learned information to put images in context.

To begin with, there is spatial context. Astronomical objects may exist in a vacuum in a strictly literal sense, but human observers learned long ago that knowledge of where these objects are with respect to one another can be at least as informative as knowledge of what they are. Spatial relationships are equally informative and important when it comes to microbes, and to human bodies. Human physicians learn the fundamentals of anatomy before they learn the particulars of disease. They first learn what and where a prostate is before learning that a spot in the transition zone of the prostate may signal a potential tumor. It is arguably this background knowledge that allows human physicians to pivot quickly in their differential diagnoses, and to remember distinctive cases that once surprised them. One could take a similar approach to AI-based image analysis, using medical image foundation models as a robust basis for diverse diagnostic objectives rather than training a new network for each narrow task.

Then there is temporal context. Whether any given object in a scene of interest is viewed as "normal" or "abnormal" depends not only on the nature of the object but also on whether it has been there before. This may be difficult to ascertain in just one imaging session. It may instead require extended observation over time. Rather than focusing on gathering complete sets of views or projections in a single imaging session, we can think about connecting different imaging sessions, and gathering only the data we need to determine what has changed.

Here is a concrete example from the arena of medical imaging. When you come in for an MRI, the machine typically collects data in more or less the same way whether it is your first, your tenth, or your hundredth time in the scanner. The particular set of contrasts the doctor orders may vary depending on your current reason for being imaged, and once new images are available the radiologist may pull up old scans for comparison. No information about prior scans is used while the new images are being gathered, however. The same basic set of projections is collected, regardless of what may already be known about you. For years, this made perfect sense to me, given what I knew about the basic requirements of tomography. Now I'm beginning to wonder. The continuous imaging approaches described in the last chapter prove that information from distinct projections obtained at different times can be combined effectively. Can't we get by with fewer projections, or different projections, if we already have lots of previous image information, and if we are interested only in what has changed?

For that matter, could an intelligent MRI machine be trained to spit out pictures that specifically highlight changes? Change mapping is already a thriving field of study for satellite imagery, which can tell us at a glance about deforestation, urban development, or strategic troop movements. Why should we not take a similar approach in medical imaging, aided by AI tools that are already disrupting the field of satellite surveillance?

One good reason for tailoring our medical imaging data to detect change is that, in many cases, change is what radiologists and other physicians are looking for. Many conditions of concern can be ruled out if one knows that an imaging finding has held steady for the last year or more. Once disease has been detected, the whole point of medical monitoring is to give warning when that disease stops holding steady and begins to progress.

Believe it or not, one trick medical imagers sometimes use to spot subtle changes in images obtained at different times is to toggle rapidly back and forth between them. This is because, in addition to highlighting edges, the visual centers in our brains are exquisitely sensitive to motion. The evolutionary reasons for this are fairly obvious—it is more important to spot a moving predator, or perhaps moving prey, than it is to get an instant fix on a tree that's been standing in place for longer than you've been alive. Imagers interested in change can exploit this built-in motion sensitivity by creating artificial motion, making it appear as though gradual changes in image appearance were happening before our eyes in real time. This is a little like what our eyes do when they jump around, creating distinctive signatures in time that help us to recognize subtle spatial features.

Questions of medical monitoring and dynamic image presentation, meanwhile, naturally call our attention once again to the axis of time. As we go through life, time is quite as important to us as space. In chapter 9, I discussed new approaches to time in imaging, modeled on the way our senses manage continuous streams of data. Now, as we contemplate linking multiple imaging sessions together, a similar principle applies, even though we are talking about different timescales—day, months, or years rather than milliseconds. We are still talking about using diverse information distributed across time to get a fuller picture of what we want to see. This kind of attunement to temporal relatedness calls to mind another deep function of our brains: namely, memory.

Memory is more than just storage capacity as we have come to understand it in the computer age. Memory is what allows us to compare old states of affairs with new ones. Great minds across the ages, from literary giants to leading neuroscientists, have grappled with the question of how memory works. Computer scientists have also dipped their toes in, emulating some rudimentary aspects of biological memory in structures like recurrent neural networks. In fact, it can be argued that training any neural network imparts a kind of selective memory. The network "remembers" key features that it has been taught to highlight, and it conveniently "forgets" other uninteresting features. Unlike a photographic memory, in which every detail is stored in its proper place, a neural network remembers the distilled essence of a thing, which with some effort can be decoded into something close to the original. In other words, learned representations are the neural network's equivalent of memories.

We all have a huge store of memories that travel with us throughout our lives. We refer back to them, update them, and sometimes add a little rose-colored tint. One particularly important way that we use memories is to predict the future. We compare current events to previous ones, and we use what has happened in the past to guess what will happen next. Whether we are conscious of it or not, we do this continually. Our brains are well-oiled prediction machines.

Take my wife, Sarah. She is always thinking ahead. Sometimes she wakes up in the middle of the night thinking about all the possible ways the coming day, or week, or month might go. The calendar in her head is so detailed as to defy proper neuroscientific explanation. I, on the other hand, am not nearly so foresightful. We joke that my relationship with time is as fluid as hers is precise. Nevertheless, my brain is still a flurry of prediction on levels ranging from the sublime to the ridiculous. Will humans ever settle other planets? When will my next chapter be finished? How can I chart a course through the seething flux of pedestrians that is Grand Central Terminal? Should I step into that intersection now or wait for the bus to pass? Where shall I place my foot so as not to trip on the curb?

From an imaging point of view, prediction is something we have more or less left to our biological brains. We make artificial images to document how things are, then we look at them and synthesize a radiology report or a scientific paper. AI is starting to change that. Many people are familiar with dynamic images on news feeds that document the potential paths of hurricanes. This is a form of predictive imaging. AI is now being used to shortcut some of the heavy-duty meteorological simulations that have traditionally been used to predict the paths of storms. Neural networks are learning to recognize weather patterns and project them forward in time.[22]

What we want with weather is early warning, and that is also what we want with health. We want to know whether we are going to get wet, and we want to know whether we are going to get sick. We want to know in advance when we are on a dangerous path, so that we can intervene before disaster strikes. I maintain that there is a way to provide advanced warning using imaging, without having to flip between images to fool our brains into seeing changes—and also without having to simulate all the innards of individual bodies the way we simulate hurricanes. We can store a series of distilled representations over time to establish a trajectory. We can learn from experience where that trajectory might lead.

This, then, is where we have arrived in our consideration of how AI may transform imaging. Instead of assembling image data into well-ordered stacks of slices in space and time for sequential inspection, as we do now, we can instead distill it down to its informative essence: a parsimonious set of highlights, distinctive features, or spatiotemporal signatures. We can funnel the deluge of data down into a manageable stream of moving representations and use those representations to home in on what is changing over time. Using whatever data is available, we can update our understanding of what has happened in the world around us—or within us—since last we checked. Then we can try to estimate what is most likely to happen next. Though this may be close to what goes on in our heads, it is very different from what we are used to when it comes to imaging.

In arriving at this model for coordinated, artificially intelligent imaging, we have also come full circle, right back to our ever-streaming senses. Intelligent machines may be just what the doctor ordered when it comes to building streams of representations that make sense of streams of data. This may be true whether the timescales of the data streams are measured in milliseconds or in years. When it comes to connecting many different kinds of data streams, moreover, the world models provided by AI may also be just the thing. Not only do we know what a cow looks like, but we also know what it sounds like. Happily or not, we know how it may smell, too. Likewise, many different types of sensors may be pressed into service to update representations initially formed from image data. If we are looking to nature as our guide for the future of imaging, then eyes are just the beginning. Connect them to a central nervous system, and support them with a sensor-rich peripheral nervous system, and we are really starting to cook with gas.

AI is a powerful tool that can be put to many different uses. Here, we have explored some new uses—image enhancement, change detection, coordinated monitoring and prediction—that may help to redefine the way we use imaging. Since these speculations take many hints from the way biological organisms see, perhaps it makes sense to ask one last fundamental question before we move on: What is the purpose of seeing?

You may legitimately object that seeing has no singular purpose. It brings us information, guides our actions, feeds our curiosity, and validates or disproves our assumptions. From an evolutionary point of view, one foundational purpose was to allow us to navigate successfully through the world. The same is true for all other visual creatures, regardless of how they see.

Once again, I invite you to consider the bat. What are a bat's goals? First and foremost, a bat aims to survive and reproduce. It strives to maintain and propagate itself, like any organism descended from the inhabitants of Earth's early oceans. Once survival is assured, it aims to thrive. To find nourishment. To hunt.

A hunting bat is a powerful model of what imaging could be. It is small and sleek, self-supervised, goal oriented but also rapidly adaptive. It searches for prey the way we might one day search for tumors or habitable planets. It rapidly distills ever-changing streams of sensory data and uses the resulting representations to plot its next move in midair. Each echo it detects may not be sufficient to generate an orderly image of its surroundings, but the complex soundscape it creates over time gives it an exquisite grasp of what is happening in the world.

We humans have not yet assembled the technology to do what a bat does the way a bat does it. All the elements are there, though. When we do manage to get our act together, we will be able to make our artificial vision not only more powerful but also more natural. In fact, the right combination of hardware and software will allow imaging to come home. I do not just mean home in the figurative sense of returning to our biological origins as sensing and predicting creatures. I mean home to our actual homes.

# 11

## No More Tunnel Vision

### *Imaging for Everyone, Everywhere*

How imaging will come home, wherever home may be.

You are lying in your bed in a comfortably darkened room. Snug and warm around your midsection, cooler where your head is exposed, you sit up, and a soft light glows on your bed table. Your bedclothes have conferred with your mattress, and your mattress has cross-checked with the digital archives of your home. Something has changed. You are not quite yourself today, but with a little help, you will be soon.

In the future, imaging will not be just a diagnostic tool of last resort. It will not be the expensive test that your doctor orders when you show up with suggestive symptoms. Instead, it will be a watchdog and constant companion that sends you to the doctor before symptoms have time to develop.

Medical imaging in the future will not live only in expensive tubes and well-resourced hospital centers. It will live with you and travel where you travel. It will be stitched into the fabric of your daily life, just as your smartphone is today. Yes, you may still be asked to lie in a fancy tube when it is time for a comprehensive checkup, or when there is some concrete reason for concern. In between such visits, though, you will be able to sit in a plush chair at your place of work or in your local pharmacy to check whether your risk of prostate or uterine cancer has changed. At home, when you retire for the evening, your bed will run through routine checks for liver anomalies. The smart bra you wear will monitor for early signs of breast cancer, and your hat will keep an eye out for changes in the health of your brain.

Many well-informed pundits have predicted that the future of medicine lies in the continuous monitoring of health—and I believe them. In *Deep Medicine*, Dr. Topol envisions a world where routine hospital beds are obsolete, patients are monitored remotely, and virtual medical assistants keep a watchful eye on our health wherever we go. My own personal physician, Dr. Devin Mann at NYU, is a believer in the future of continuous care, and he leads a team that is working to speed us on our way to such a future. Devin divides the more than three-thousand-year evolution of health care into three stages: the premodern era of house calls, in which caregivers visited people in their homes; the modern era of episodic care, in which patients instead visit their caregivers in centralized health care facilities; and a future in which people will benefit from continuous care. Patients in the future, he argues, will once again receive much of their care in their homes, backed up as needed by episodic care in ambulatory or hospital facilities.

In my own lectures on the future of imaging, I have taken to using a summary slide of Devin's that outlines this broad trend in health care delivery. I generally go on to point out a key problem for imagers like me: imaging that resides in ambulatory or hospital facilities—which is to say nearly all medical imaging at the present time—would by definition be a backup player in a future of continuous care. The true heroes of continuous care would be home monitoring devices: smartwatches to flag irregular heartbeats; smart bathrooms to pull evidence of metabolic irregularities or infectious exposures from our daily ablutions; smart apps to detect concerning departures from our accustomed browsing histories or movement patterns that might indicate changes to our mental or physical health.[1] Expensive imaging machines, and expensive imaging specialists, would wait in the wings, called up only when needed to sort out anything the monitors can't handle. People would be healthier in such a future, and, however disappointing it might be to be relegated to the second string of new high-tech health care teams, my imaging colleagues and I would learn to carry on.

Here's the rub. Unlike some current proponents of smartwatch-driven health care, I do not believe that wearable or environmental or even chemical and genetic sensors will be able to do all the heavy lifting on their own. As an imaging physicist, I have learned through long experience that any individual sensing mechanism has stringent limits on how far and how clearly it can see into the body. It is no great stretch for an Apple Watch to record your EKG, whose electromagnetic signature is distributed through-out the body, but ask that watch to see more than a few centimeters into your wrist, and it quickly goes blind. Chemistry can be a potent probe of pathologic changes occurring at the cellular level, but chemistry alone is not very good at telling precisely where any cells that are going off-script might actually reside. Our bodies are more than just collections of signals or bags of cellular secretions. How we are put together is a significant part of who we are, and disease often manifests as something inside us being out of place. When it comes to health, then, telling here from there can be a pretty powerful advantage, as nature taught us long ago.

Long ago, natural organisms learned how to place sensory information in context, weaving diverse incoming signals into composite spatial maps of the world outside. How much more powerful would our futuristic sen-sors be if they were similarly grounded in detailed maps of inner space? How much more powerful still if those spatial maps were tied together over time, so that changes of concern would leap out at us like zebra stripes?

To accomplish this—to put imaging at the center rather than on the fringes of continuous care—we will need to liberate imaging from its tubes. Even if modern medical imaging protocols are starting to become more continuous, imaging sessions are still comparatively few and far between. The race is already on to fill geographical and socioeconomic gaps in access to imaging devices. A new approach to imaging, and to image-derived information, will also be essential to fill yawning gaps in time and deliver on the promise of continuous care.

This chapter will briefly summarize what it is that has kept medical imaging bound up in tubes. Then it will invite you to consider how we can escape the confines of those tubes in order to fill key gaps in space and time. The chapter will conclude by panning out from the medical per-spective to explore how other high-tech imaging technologies will come down from the mountaintop and make their way into local neighborhoods around the world.

Image quality is a harsh mistress.

Progress in imaging is often measured by advances in image quality. Over and over again, across the entire history of human-made imaging, our images have gotten visibly and measurably sharper: they look crisper to our eyes, and the information they convey is ever more precise. It is as if the entire arc of imaging history has been an exercise in focusing.

The notable exception to this trend occurs whenever a new imaging modality is first introduced. Imaging pioneers get a pass when they are unveiling something unexpected, and people squint appreciatively at even crude-looking offerings like Röntgen's first x-ray shadow pictures. Then the recurring process of refinement begins. Astronomers start out with fuzzy blobs in the sky, which are pulled into sharp focus as time marches on. Radiologists rush to make divinations from blurry images, until imaging vendors put new high-definition hardware in their hands.

Once a new frontier of image quality has been reached, however, it is ever so difficult to go back. Radiologists today complain about faint artifacts in images that would have knocked the socks off previous generations. In some situations, it has been shown that radiologists demand a good bit more image quality than they actually need to make particular clinical calls. And why wouldn't they? If you were making life-or-death decisions, wouldn't you want nothing less than the best?

There is some interesting neuroscience to all of this. It turns out that the brain is extraordinarily good at remembering images it has encountered before. Neuroscientists like Professor Nicole Rust of the University of Pennsylvania have studied the process by which humans and primates identify images as new or familiar. Their studies show that a sense of familiarity is remarkably hard to shake. The hypothetical mechanisms by which human or primate brains accomplish the recognition of sameness or difference are fascinating, with some provocative analogies to (and doubtless many lessons for) artificial self-supervised learning. The bottom line is that once you have seen an image, it is hard to unsee it. Likewise, I would posit from experience (though admittedly without the same rigor of neuropsychological testing to back me up) that once human specialists have gotten used to top-quality images, it is hard for them to be satisfied with pale imitations.

I maintain that it is largely the demands of image quality that have kept medical imaging devices tied to big tubes, resisting the prevailing technological winds of miniaturization. Principles of tomography dictate that image quality must be purchased with a requisite number of carefully calibrated and coordinated projections. In order to reconstruct the cross sections they need, imagers must know exactly where their detectors are pointed, and they must be able to point them where they choose. They must home in carefully on signal, and just as carefully tune out noise. This takes time, expertise, and engineering. It also gives imagers tunnel vision.

The psychological imperative to maximize image quality presents a formidable challenge to pioneers of the less-is-more movement in imaging. Indeed, numerous previous attempts by imaging vendors to develop low-definition or limited-use devices have stumbled in the market, for reasons that would make perfect sense to any computer manufacturer. People who don't already have computers might purchase outdated models at a discount, but who would want to buy a new machine whose performance was worse by design, if the tasks for which they needed a computer were the same as they had ever been? This is one aspect of what has been called the innovator's dilemma.

The innovator's dilemma, as spelled out in a well-known book of that name, explains why market leaders are seldom able to disrupt their own markets, even if they got to their position of prominence in the first place by developing disruptive technologies.[2] Disruptive technologies, so the reasoning goes, are typically commercialized first in emerging or insignificant markets. However, successful companies quickly become accustomed to meeting the demands of their best customers, who may not think they want new disruptive products. Market research only cements these preconceptions. By the time managers in market-leading companies are able to make the case for new low-margin products, it is often too late. Smaller, more nimble companies who are less tied to maintaining market share are already on the way to eating the established companies' lunch. The innovator's dilemma goes a long way toward explaining why big imaging vendors' previous efforts to buck the trend of progressive image quality improvement have by and large failed. For all their success in turning imaging into big business, these companies have not excelled at building up fringe markets to the exclusion of the established needs of their principal customers, like radiology departments, cardiology practices, and hospital systems.

The new generation of medical imaging innovators is taking a different tack, both in academia and in various hot new start-ups that are popping up left and right nowadays. These would-be disruptors are aiming their less expensive, more accessible devices squarely at new markets, whether those be historically imaging-poor regions of the world, or else point-of-care settings like intensive care units or mobile emergency services which have traditionally been inhospitable to bulky imaging devices. Even the new imaging innovators, though, are not immune to the siren song of image quality. When they tout their new systems, they still commonly compare the resulting images with images from more expensive state-of-the-art imaging equipment. They are still intent to show that, even if their systems do not yield leading-edge image quality, they still produce images of far better quality than one would have thought possible for the price. By accepting the familiar metrics by which traditional imaging is judged, they have unwittingly submitted to the tyranny of image quality.

Make no mistake: image quality *is* important. If you haven't given careful thought to optimizing quality, then you probably haven't done your job as a designer. The image quality metrics that are commonly used for optimization, though far from perfect, are relatively good at assessing the general degree of spatial discrimination and the overall preservation of image contrast. The problem is that if you optimize *only* for image quality, you will likely end up right back at current system designs, which have undergone decades of development to eke out every available bit of quality. If you measure yourself as others have always done, then it is hard to truly break the mold.

Image quality is important, but it isn't everything. Radiologists can make important calls from limited images, and so can AI. While stereo vision gives us a flimsy picture of depth that does not always hold up under rigorous scrutiny, it still delivers actionable information that allows organisms like us to survive and thrive. As these examples illustrate, *information* quality is ultimately more fundamental than image quality. What is an image, after all, but spatially organized information? The challenge with using information content as a metric of quality is that the results depend on what information you want, and what you want it for. That is where things can get really interesting in the future.

Chapter 10 explored various emerging artificially intelligent technologies and transformations that might propel imaging forward for untraditional use cases. We discussed representation learning, self-supervised learning, and the idea of tracking distilled representations over time. These are tools that can be used to jailbreak future imaging devices.

Let us now consider the possible shape of tomorrow's AI-powered medical imaging devices. To begin with, in addition to a brain, smart medical scanners will gain a memory. Previous images and other relevant information will be used to tailor the new data that is gathered, making imaging sessions progressively faster, and focusing them explicitly on detecting departures from your baseline state of health. One of the reasons pancreatic cancer is so deadly is that it grows gradually under the radar, and it is generally not detected until it has already started to spread. Imaging machines that are specifically tuned to detect small changes over time could provide the early warning that is critical for survival. Prostate cancer, by contrast, can percolate uneventfully for years or decades after it is discovered. The key is to detect when it starts to become aggressive and life-threatening. Here, MRI has already been shown to allow noninvasive monitoring in place of regular invasive biopsies, and change-attuned machines could make longitudinal monitoring faster, cheaper, and more accessible. That is why my NYU colleague Hersh Chandarana—an expert in imaging of the abdomen and pelvis—is obsessed with scanners of the future. Patricia Johnson, a physicist in our radiology department, is working with Hersh and his abdominal imaging colleague Angela Tong to make scanners intelligent enough to decide automatically after just a few minutes whether a patient is at high enough risk of prostate cancer to continue scanning. My graduate student, Arda Atalik, is working on emulating depth perception by using contextual information derived from prior scans or text reports to enable ultra-fast imaging with small numbers of projections. And Lavanya Umapathy, a postdoctoral fellow in my laboratory, is exploring how giving neural networks memory of a patient's prior images can make the networks better at estimating disease risk by enabling them to detect which features of current scans are new and which have remained stable over time. The longer-term goal is to change the way scanners are operated, and even the way they are constructed.

Image quality for the first generation of new memory-enabled imaging machines—let's call them type 1 scanners of the future—will likely equal or exceed the current state of the art. The machines themselves will likely resemble today's scanners in their overall bulk and expense. However, their operation will be both newly efficient and newly data-rich. They will be bristling with peripheral sensors that will track in detail each patient's movements, as well as the machines' own internal states. They will gather data in continuous, multisensory streams, adapting on the fly as conditions shift. They will be the medical equivalent of self-driving cars, always monitoring inside and out while hunting for worrisome changes that might affect their precious cargo.

In the current episodic model of medical imaging, the task of arranging distinct episodes into a storyline is left largely to the mind of the physician. Soon, though, our scanners will be able to tell their own stories. They themselves will be able to sound the alarm at any concerning changes. Are you on a slow trajectory toward Alzheimer's disease? Has there been an ill-defined but clearly uncharacteristic shift in your body's structures since the last time you were scanned? These are the kinds of questions that smart, adaptive, and predictive imaging could answer.

Eventually, memory and predictive power will allow the design of medical scanners themselves to change, making them smaller, cheaper, and more accessible, and delivering in earnest on the promises of the current less-is-more trend. When image quality inevitably suffers, as the physics of tomography dictates that it must when certain engineering compromises are made, the use of image memory, representations, and correlated sensor streams will help to rescue degraded images. Perhaps more importantly, though, these coordinated tools will also allow us not to focus so single-mindedly on producing visually pleasing images. In a revolutionary break from the tyranny of image quality that has defined the progress of imaging for millennia, new imaging devices will arise with the specific purpose of providing early warning, as our own senses evolved to do for us. Cheap scanners will be deployed in doctor's offices, or in drugstores. Rather than producing images that require advanced training to interpret, they will provide simple answers, like any other medical test. They will tell you if you are continuing in your baseline state of health, or if instead you might need to go in for some more advanced imaging. Think of these cheap, answer-oriented devices as type 2 scanners of the future.

But why stop at the drugstore? I fully expect that the latticework of sensors built into future scanners will eventually escape the bounds of our time-honored tubes entirely. If sensors can be trained to operate like auxiliary senses within high-end medical imaging devices, then why can they not also operate on their own, informed by memory of previous images? Cheap sensors—type 3 scanners of the future—may not be able to produce state-of-the-art cross-sectional images in a traditional tomographic sense, but they may still be able to provide early warning of changes to your health.

Consider this analogy. Once you have carefully inspected the layout of a room—once you have built up a sufficiently robust representation of it in your mind—you can generally close your eyes and navigate pretty effectively by touch, sound, and other senses. You may not be quite as agile as you were when you had your eyes open, but you can get by. At the very least, you should be able to sense when someone new enters the room. Now invert the scene. Instead of you training your senses on the room around you, imagine that the room is focusing attention on you. Imagine that the walls, the furniture, even your clothes are instrumented with a diverse network of sensors with built-in spatial awareness. Imagine that all the room's senses are on the alert for internal changes, collectively attuned to signatures of approaching disease. You can also think of type 3 scanners of the future as a new twist on self-driving cars: not smart vehicles steering you through a changing environment, but smart environments looking out for concerning changes in you.

Unlike today's imaging infrastructure, the technology required would be low profile, lightweight, and inexpensive, with no precision-engineered tubes required, and no interpretation by highly paid specialists needed either (except, of course, when your early warning system has been activated, at which point smart high-definition scanners and experienced radiologists would pick up the ball again). All hope of generating crisp images from type 3 sensors would not be entirely lost, either. Rather than marshaling precise and carefully ordered projections, one could use AI to allow cheap and messy projections to stand in for clean, expensive ones.

A few years ago, my imaging colleague Bruno Madore and his coworkers at the Brigham and Women's Hospital in Boston started placing little ultrasound probes on subjects inside MRI machines. They weren't even looking to generate ultrasound images; they just sent unfocused pings into the body and recorded the jumble of echoes that resulted as the pings bounced back and forth off of internal structures. They then treated each distinct jumble of echoes as a signature—a kind of fingerprint. They associated every ultrasound fingerprint with an MR image gathered at around the same time. Then, they took the subjects out of the MR scanner and continued to make ultrasound measurements. By finding the closest match for each new ultrasound fingerprint, they were able to call up corresponding high-resolution MR images in real time, without gathering any new MRI data. They called this feat scannerless real-time imaging.[3]

Others in the MRI field had questions. How robust were the associations between ultrasound pings and MR images? Which image details could be trusted, and for how long after the initial MR scans? Could any new image information be discovered, or was scannerless imaging limited to the exact set of MR images used for training? These are all legitimate questions, worthy of study. You can see why the prospect of scannerless imaging might have caught my eye, though. It is not too great a stretch to imagine learning shared representations that capture the mutual information between ultrasound fingerprints and MR images, then using these representations to decode ultrasound fingerprints into at least approximate high-resolution images. Provocative proposals for something like indirect tomography using probes other than ultrasound are beginning to appear. For example, optical imaging researchers have trained neural networks to convert complex time signatures of the scattering of red light into crisp images of blood vessels, and my colleague Leeor Alon and his team have converted electrical signals from a microwave helmet into cross-sectional brain images resembling MRI.[4] In principle, any penetrating probe like radar or infrared light could also be used to find fingerprints that correspond to detailed cross-sectional images.

But I digress. Do you see how tempting the lure of high-quality images can be for an imager? As we have already established, scanners of the future may not be single-mindedly focused on the generation of images per se. Instead, the principal value of such scanners may lie in establishing robust individualized representations of health and then tracking and updating those representations using all available means.

This brings me to the *Star Trek* tricorder. For those of you who have not seen reruns of the campy but visionary *Star Trek* television series or its many subsequent spin-offs, the tricorder is a handheld device that, when waved suggestively near an ailing subject, instantly diagnoses the problem (and sometimes even fixes it). As silly as it may sound, the tricorder has long served as a kind of holy grail for imaging scientists. It does everything we aim to do, without all the fuss. While *Star Trek*'s creators never quite let on what futuristic probes the tricorder used to gather its internal information, the device they envisioned was conveniently portable, appealingly versatile, and tantalizingly infallible. For years, professional fans like me have been left with the pesky question of how, in actual real life, to accomplish such a marvel.

I am coming to believe that the tricorder's visible wizardry might be a bit of misdirection worthy of Oz. If we were to travel into *Star Trek*'s high-tech future and pull back the curtains, I suspect we would see that the little handheld device is just the tip of the iceberg. I suspect that we would discover it is merely a generator of cheap projections—a finder of suggestive fingerprints, backed up by banks of digital representations elsewhere on the Starship Enterprise, and by an even more immense store of information amassed over time by all the disparate members of the United Federation of Planets. Rather than rewriting the rules of tomography entirely, the tricorder might just use intelligent transformations to fine-tune all this collective learning to each individual patient, identifying anomalies in the patient's current data, comparing the patient's recent health trajectory with the trajectories of other patients in its records, and making predictions about the patient's future state of health. Such a concept is not entirely futuristic. It is a little bit like using world models built up over time to add depth to cheap two-dimensional projections. Perhaps the concept of the tricorder doesn't belong only in a distant and visionary future. Perhaps it is just another familiar form of vision.

As is illustrated in figure 11.1, the scanners of the future that I have sketched for you so far represent multiple interconnected tiers of imaging technology. Advanced cross-sectional imaging devices in hospital or outpatient settings—that is, type 1 scanners with memory—would establish a detailed baseline representation for each person, ingesting diverse information from the electronic health record to flesh out the picture, performing fine adjustments at intervals, and hunting for faint signs of encroaching disease. Cheap, answer-oriented type 2 scanners, built into plush chairs or examination tables in easily accessible settings like drugstores or places of work, would fill in the gaps between advanced imaging sessions, performing coarse adjustments to shared representations and raising warning flags when concerning changes are detected. Tabletop, wall-mounted, or wearable type 3 sensors integrated seamlessly into your daily life would make more continuous updates to shared representations, once again flagging changes that might merit further investigation. All of these medical monitoring technologies would be tied together by running multisensory representations.[5] Medical foundation models would be trained to combine features from imaging and sensing data with features from blood tests and genetic surveys to create still more informative representations. Self-supervised learning would be used to tell if today's representation is significantly different from yesterday's, or if the signal from the ultrasound sensors in your belt corresponds to the same state of health

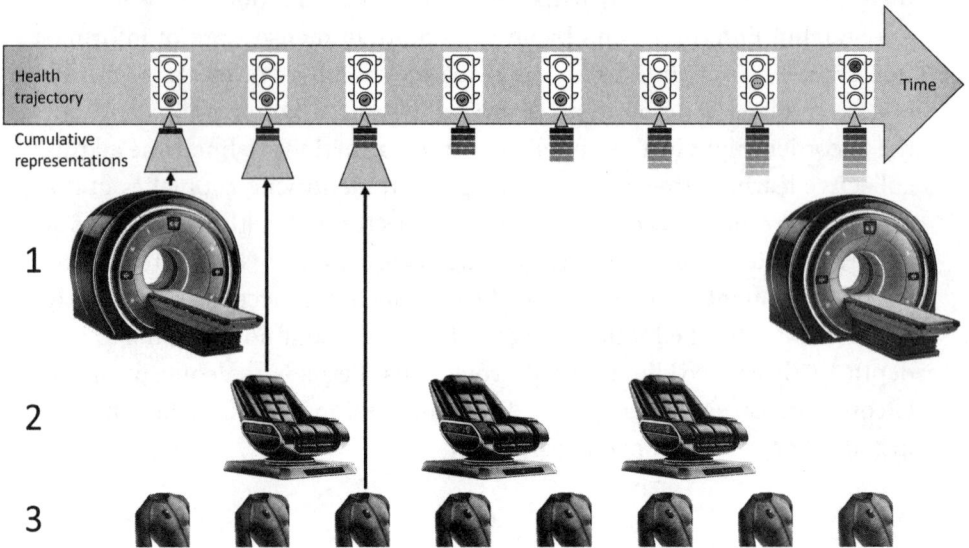

**FIGURE 11.1** Everywhere scanners: multiple tiers of imaging technology connected by AI.

*Source*: Created by the author with DALL-E 3 (OpenAI) and Microsoft PowerPoint.

you exhibited at your last MRI scan. Data gathered over time in large populations would be used to identify subtle trajectories toward disease that might be difficult to see in a single individual's data. Like our everyday lives, our medical care would be self-supervised but also informed by all the people who have gone before us.

The name I have come to use for this tiered set of interlinked technologies is "everywhere scanners." Unlike modern high-maintenance scanners, their reach could extend, well, everywhere. Just as importantly, their use would not be limited to carefully choreographed episodes. They would allow medical monitoring not just everywhere but all the time. They would no longer be the last bastion of diagnosis, called into service when other more convenient but less definitive means of testing have been exhausted. Instead, they would be on the front line of continuous care. They would be our individual and collective early warning system.

Consider the potential impact of such an early warning system on one particularly dreaded disease: cancer. In 2022, the American Cancer Society estimated that forty percent of the world's population (i.e., three billion people at the time of writing) will be diagnosed with cancer. Fifty percent of that unfortunate community (namely, 1.5 billion people) will receive their diagnosis late. The overall twenty percent survival rate five years out from diagnosis would be increased to eighty percent if the cancer were detected early. That's a total of nearly one billion living souls who would otherwise have been lost.[6] As incentives for technology development go, a billion lives is a humdinger.

A friend of mine named Emi Gal summarizes these stark statistics in a particularly pithy way. "We already have a cure for cancer," he says. "It is early detection." Emi, a Romanian coding whiz turned serial entrepreneur, is the CEO of a company called Ezra, whose mission is to catch cancer early in as many people as possible. In the service of this mission, Ezra offers full-body MRI scans to its customers at regular intervals. Such a practice flies in the face of standard radiologic guidelines, at least for people who do not already have a documented high risk of cancer . . . and that is precisely why it interests me.[7]

Mention the word *screening* in professional medical imaging circles, and you might feel a chill descend upon the room. This is because population-wide screening can be surprisingly difficult to do well. Numerous sources of data show that indiscriminate screening leads to expensive but poor-quality care—what Eric Topol calls shallow medicine.[8] The reason boils down to simple statistics. No screening test is perfect, and when the probability of disease is low, the number of false-positive test results will far exceed the number of true positives. The risk associated with false-positive results will then outweigh the benefit of true-positive findings, at least at a population level. If you take a random person off the street and trundle them into a scanner, for example, you are bound to see more than you bargained for. You will discover all kinds of incidental findings, most of which will be red herrings. The obligatory follow-up for these false positives will do more harm than good, resulting in unnecessary tests, unwarranted treatments, unjustified expenditures, and pointless anxiety.

In the second chapter of *Deep Medicine*, Topol provides three cautionary tales to illustrate the downsides of screening. First, there is the case of mammography, which has long been touted as a staple of breast cancer prevention, and which generates about $10 billion of cost per year in the United States alone. According to the statistics Topol cites, sixty percent of people undergoing mammography receive at least one false-positive result, leading to unnecessary biopsies, surgery, radiation, or chemotherapy. Only

0.05 percent, however, avoid death from breast cancer. A notable correlate may be found in prostate cancer screening using prostate-specific antigen (PSA), which is still used for thirty million people in the United States each year, despite a recent change in clinical practice recommendations by medical advisory bodies. Twelve to twenty-four percent of those tested receive false-positive results from PSA screening, and four to eight percent undergo unnecessary radiation therapy or surgery, whereas only 0.1 percent avoid death from prostate cancer. Then there is the case of thyroid cancer screening using ultrasound and other tests. When concerted thyroid screening programs were tried, both in South Korea and in the United States, the apparent incidence of thyroid cancer skyrocketed, since it was now being detected regularly, but the programs resulted in no measurable change in mortality from thyroid cancer.

Indeed, early diagnosis alone, Topol argues, may not always improve outcomes as much as we once thought. Some cancers are now known to metastasize far earlier than was once believed. The value of early diagnosis becomes difficult to assess, moreover, when effective therapies don't exist, or when available therapies are themselves high risk. Let me be clear: appropriate screening is essential for people with known risk of disease. For such people, the benefits of true-positive tests far outweigh the risks of false positives. It is what Topol calls promiscuous screening that is to be avoided.

In light of all this information, how is the deployment of everywhere scanners to be justified? Wouldn't the use of such devices in the general population lead to fountains of false positives? Not, I would argue, if we design them right. An effective early warning system must be designed not for one-shot screening—like a traditional diagnostic imaging examination—but for longitudinal monitoring, like recurring Ezra examinations. At first blush, screening and monitoring may sound the same, but they are not. Screening results are often interpreted in isolation, and normal values are generally based on population norms. They are not individualized, and they do not really take advantage of the axis of time. Monitoring, on the other hand, follows an individual subject's trajectory, and it is specifically focused on change over time.

With memory, change tracking, and predictive power on board, everywhere scanners would transform the statistics of early warning. Radiologists today already know that many things seen incidentally on isolated scans can instantly be ruled out as items of concern if they appear unchanged on previous scans. This is why advanced tomographic imaging like MRI is widely accepted nowadays for periodic monitoring in people who are known to be at risk for breast or prostate cancer—two of the populations that have arguably been less than ideally served by shallow screening. Imagine how much more confident patients and physicians could be with a robust timeline of images, sensor data, and corresponding representations to work from. Neural networks could also be trained specifically to rule out false positives. If you are tempted to complain that these networks might then start missing important findings, remember that any reliable information is better than none at all. In the current paradigm of episodic care, precisely nothing is done in the intervals between episodes, so even excessively conservative information, if attainable at low cost and high convenience, would be a win. Meanwhile, imagine the benefits of all the robust information we would gain from population-wide monitoring when it comes to fleshing out our collective understanding of disease evolution. Imagine what biobank-minded scientists like Karla Miller could do with worldwide repositories of high-frequency data. They could build detailed maps of the complex dynamics of inner space.

The prospect of dynamic image-informed representations calls to mind another concept much touted by today's scientists and futurists. I am referring to the notion of a "digital twin." This is often conceived of as a top-to-bottom simulation of everything that makes you you—a kind of digital mirror image that follows you through space and time, modeling everything that happens inside you from your cells on up. In the context of personalized health care, one could run the simulations forward in time and raise the alarm whenever a future you gets sick. While I love this idea in general, I would certainly balk if I were asked to build such a digital twin with twenty-first-century technologies. I would find the task at least as daunting as the task of constructing a practical tricorder. That said, it is possible that digital twins need be no more magical than tricorders once we step back from the hype. Maybe dynamic representations are all we need: not ground-up simulations of mind-numbingly complex ensembles of interacting cells, but top-down distillations of the various informative imaging and sensing signals we have at our disposal.[9]

Physicists have dealt with the problem of intractable complexity by creating what are called effective theories. Rather than trying to track every subatomic particle that makes up the bulk of ordinary matter, they use ingenious mathematical strategies to sum over multiplicities of internal states and derive succinctly defined average behaviors that apply at any chosen scale in space and time. Interestingly, some of the averaging strategies used by modern physicists have been shown to have deep connections with the distillation functions performed by artificial neural networks. From this point of view, distilled representations of health could be viewed as "effective digital twins." Their function would not be to mimic your every move. Instead, they would be sparse but distinctive sketches, encapsulating key changes in you that might call for more careful inspection. They would not be simulacra, but rather digital guardians, charged with keeping a watchful eye on your welfare. They would not copy you but rather connect you with yourself over time.

What will happen to radiographers and radiologists in an era of everywhere scanners and roving digital guardians? Will these imaging specialists finally fall prey to Hinton's prediction and be replaced by smart machines? I don't actually think so. At least for some time yet, people will still be called upon to figure out exactly what is going on when an early-warning flag is raised. In fact, far from overturning today's imaging paradigms, tomorrow's disruptive imaging technologies may actually *expand* them. Type 2 and type 3 everywhere scanners will be a feeder, not be a substitute, for comprehensive radiologic surveys. Radiology departments equipped with type 1 scanners will see an influx of brand-new cases caught by the continuous-care safety net. These cases will be just the sort of cases radiologists prefer: not a flood of negative studies to sort through exhaustively, but a prescreened set of high-suspicion scenarios in which a specialist's expertise can make a real difference. In other words, experts in image acquisition and reconstruction will have their work cut out for them, and patients everywhere will reap the benefits. One day, I suppose, high-performance scanners may learn to navigate complex cases well enough on their own, and neural networks may gain enough intelligence to make their own uncontestable recommendations for medical management. By that time, I hope that today's human imaging experts will have moved on to do amazing new things.

Before we leave the topic of health care, let me note that new modes of imaging will not just improve health through early warning. They will likely transform medical treatments as well. Image-guided therapies have already made a huge difference in the practice of medicine. From the first days of x-rays onward, surgeons have planned and executed their procedures with help from images. When laparoscopic surgery (which is to say surgery performed through small holes under the guidance of inserted cameras) came along, it saved countless people from dangerous infections, painful recoveries, and unsightly scars. More recently, in situ imaging has begun to provide surgeons with key information about where to cut. For example, the surgeon-scientist Dr. Quyen Nguyen of UC San Diego has developed dyes, visible on heads-up displays, to highlight the margins of tumors that must be extracted from normal tissue, and to trace the path of nerves that should not be touched.[10] Nowadays, robotic surgeries are gaining ground, with imaging providing essential real-time feedback to remote human surgeons. Energetic imaging probes, meanwhile, are increasingly being used to effect cures without any cutting at all.

In chapter 4, while exploring the principles of tomography, I mentioned that similar principles could be used to focus therapeutic beams on tumors and other unhealthy tissues. My dear friend and onetime college roommate Dan Sidney proved this point to me long ago. For many years, Dan worked for a company called TomoTherapy, whose principal product was a CT-like machine that delivered carefully modulated therapeutic radiation as it spun. The net result of the radiation projected at each angle was a tailored cumulative radiation pattern, like a tomographic image imprinted directly onto tissue. Other more common radiation therapy devices dispense with the CT tube but use the same general mode of operation.

For this form of hands-off therapy, it really helps to see what you are doing, and intelligent imaging could make all the difference. Some of the continuous cross-sectional imaging approaches surveyed in chapter 9 are already being pressed into service for image-guided radiation therapy. Likewise for the emerging field of MR-guided focused ultrasound therapy, in which MRI is used to outline the target tissue and map out changes in tissue temperature as the high-intensity ultrasound does its work. Patients and doctors alike hail the results as near magical. I have seen videos: a patient with Parkinson's disease, who started the day with a debilitating tremor, gets up off the table with steady hands, a new lease on life, and a completely intact skull. Tricorder, here we come . . .

The picture I have painted here of the future of medical imaging is just one manifestation of how the broader future of imaging might play out. Other parts of the imaging portfolio are showing evidence of developments similar to those that promise to revolutionize medicine. Cheap sensors are proliferating, and they are operating more and more in tandem. Imaging sessions are becoming more continuous, and more connected. AI is on the move. More and more, our artificial sensory organs are coming to resemble our natural ones: always on, always learning, always on the lookout for change.

Just the other day, I sat in on a lecture by the computer scientist Guillermo Sapiro, who described how he had teamed up with an autism expert named Geraldine Dawson. Sapiro explained how their research team had trained artificial neural networks to analyze video clips for subtle signs of autism, tracking the way kids look at pictures and how they respond when their parents call their name. The result was an automated diagnostic tool that could substitute for hours of intensive observation by clinicians. Its use in areas with limited access to busy pediatric mental health specialists could cut years off the waiting time for definitive diagnosis. A computer that can literally see autism: how about them apples?

And how about imaging applications that have no medical angle at all? Let's talk about astronomy. By many measures, astronomy has set the standard for species-wide collaboration. It has also pioneered the use of coordinated imaging devices on a truly planetary scale. As I see it, astronomy still has many new frontiers ahead. The National Radio Astronomy Observatory operates a 256-element radio telescope array that searches the sky for transient cosmic events. Its detectors are sensitive enough to detect micro-sparks in the metal staples embedded in wooden power line poles miles away. Do you know what other devices detect radio waves, though? Every cell phone in every pocket on God's green earth. It is not too difficult to imagine a future in which all cell phones on the planet are linked together as part of one immense radio telescope. Any working radio astronomer reading this now is probably having conniptions over just how I imagine this might work in practice, but the radio astronomers I've polled so far haven't been able to come up with any fundamental reasons why it wouldn't work. It would be like the world's biggest compound eye. The signal detected in each phone might be far too small to rise above the noise, but put them all together and one could exceed the resolving power of the biggest, most imposing radio dishes in existence. Such a feat would require exquisite coordination—the equivalent of tracking billions of independently jittering eyes and synthesizing their views into a clean composite picture—but as you have heard from me repeatedly, our brains have already set us a clear precedent.[11] Add a network of space observatories like the remarkable but as of yet solitary James Webb telescope into the mix, and we would have ourselves something that would make today's most powerful telescopes look like children's toys. We would have an extraplanetary sense of sight, with data streaming in from everywhere all the time. This would truly be imaging for everyone, and, by virtue of the hardware in our pockets, it would also be imaging *by* everyone. Who knows what wonders we might see together?

So much for space, but let's not forget about time. The idea of using moving representations to connect different imaging sessions is not just a pipe dream for medical imagers like me. Astronomers and microscopists are very interested in dynamics, too. They want to trace the transits of exoplanets and catalog the cataclysmic eruptions of supernovae. They want to track the living dance of molecules and cells. Why should they not also emulate our senses and our brains in order to do so?

Speaking of astronomy and microscopy, I expect we may have yet another form of coordination in our future that brings inner space and outer space together in new ways. At the moment, it is a laborious proposition at best to stitch together images obtained at vastly different resolutions. People have certainly tried, and various multiscale atlases of tissues or of galaxies currently exist. Most of us are familiar with the canonical order-of-magnitude sequences in movies that zoom us out from the earth to the cosmos, or zoom in from the familiar world of people to the exotic world of atoms. What if we could zoom in and out like this in real life? In some ways, we are worlds away from such a capability. In other ways, we are closer than you might think. The emerging field of spatially resolved genomics allows the detailed mapping of individual cells in the midst of complex tissues, with volumes of genetic information available for each subcellular pixel. I don't imagine that Google is quite ready to expand its Google Earth tool to include "Google Body" or "Google Cell"—but stay tuned.

Recently, as I was reading the book *Poseidon's Wake* by Alastair Reynolds— a favorite author of mine and a master of the sci-fi genre of space opera—I came across the following imaging tableau, tossed out in passing in the course of a rollicking plotline: "Scanning systems orbited her like a host of tiny whirring satellites. Goma watched as the scanners slowly assembled a three-dimensional image of Eunice with sub-cellular resolution on a variety of flat and solid display media."[12] Here, Reynolds, who was trained as an astrophysicist, casually connects the macroscopic scale of a person with the microscopic scale of cells and their components, all using the metaphor of a planet and its orbiting satellites. Being the particular reader that I am, I found myself musing on how the composite image in question might be assembled—but never mind that for now. Good science fiction often poses challenges to today's scientists. Consider the gauntlet thrown.

The future of imaging does not lie solely in big machines or in small ones. In this chapter, I have tried to imagine where bigger-is-better and less-is-more might come together. I have envisioned a world where imaging is done differently: a world populated by huge collections of tiny sensors, and by networks of smart machines that look out for us. In the process, I have made much of analogies with our naturally evolved senses. This raises the question of how we might take the next step in evolution, augmenting our natural senses directly with other senses we have built. It raises the question of how imaging may once again change the way we see. Not just metaphorically, but literally. Not only for better, but potentially very much for worse. We have explored the future of imaging that might emerge from our long history of seeing. Let us close the circle now and consider together the future of seeing.

# 12

## The Future of Seeing

What imaging shows us about who we are, and what
we might become.

**W**hen I look ahead, I see a bright future. In this future, we have learned to see the world, and ourselves, in an abundance of new ways.

I also see a dark mirror. Shadowed threats lurk in murky depths, preventing us from seeing one another with any clarity at all.

Which one will it be? That is up to you.

~☙~

Omnipresent cameras, everywhere scanners tracking inner space continuously, personal communication hubs that are both individual imaging devices and interlinked components of planetary-scale telescopes: this is just a glimpse of what the future of imaging may have in store. If we take to heart the lessons of biology, if we build upon current trends in technology, and if we give the power of transformations free rein, we are bound to find brand new ways of emulating and augmenting our vision. One noteworthy outcome will be a progressive democratization of imaging. More and more, technologies that were once the exclusive domain of imaging professionals will make their way into the hands, or in any case the lives, of everyday people.

In a way, this vision of the future brings us right back to the origins of vision itself. By the time the Cambrian explosion had come to a close, there were eyes everywhere. The seas were brimming with spy cams built into the bodies of trilobites, mollusks, and chordates (the ancestors of modern fish, amphibians, reptiles, birds, and mammals). What will we do when our day-to-day environment is likewise teeming with the organs of artificially enhanced vision?

Democratizations of imaging have happened before. Photography brought artificial images into people's homes, then into their pockets. X-rays opened people's insides up to easy public inspection. For a time, bigger-is-better imaging has managed to buck this trend, but I don't think it will do so for very much longer. And when the new wave of up-close-and-personal imaging does arrive, it may come in like a tsunami.

I have already alluded to potential benefits for public health. New forms of imaging could save billions of lives by catching cancer and other deadly diseases early. Imagine the benefits for quality of life if slow, degenerative conditions that affect your aging brain or your creaky joints could also be headed off early. Imagine the benefits for health care spending, too, if inexpensive early-warning technologies could diminish the need for expensive therapies.

Beyond merely increasing the average duration and quality of life, a broad dissemination of imaging technologies could address a wide range of current inequities. For example, faster, easier medical imaging could reduce wait times and increase the accessibility of limited imaging resources. New smaller, cheaper scanners could provide imaging capability to underserved parts of the world. Cheap multisensor monitoring could fill in the gaps between more expensive imaging sessions. A capacity for cheap and continuous health care that is predictive rather than reactive would certainly help to head off many medical challenges that are currently endemic in poorer populations.

A new democratization of imaging would impact not merely health equity, but other forms of equity as well. Computational photography is already yielding more accurate visual representations of people of color, as I have mentioned, and many ongoing manifestations of racism have been called out by bystander videos and body cams. In addition to being a tool for transparency, imaging is a powerful tool for discovery. Equitable access to the tools of imaging, therefore, can promote more equitable access to knowledge. The more new telescopes there are, the more Galileos there are bound to be.

"Give a person a fish," the saying goes, "and you feed them for a day. Teach a person to fish, and you feed them for a lifetime." Real equity in imaging will include equity in the ability not only to consume images but also to generate them. Like many areas of science and engineering, the field of imaging itself has struggled historically with all manner of socio-economic, gender, and racial imbalances.[1] Imagers like Rod Pettigrew have had to break through color barriers in order to have a seat at the imaging table. Organizations devoted to the advancement of imaging have only relatively recently begun to grapple with the differential advancement of women in the field. (In 2023, the list of female editors-in-chief of medical imaging journals rose to a historic complement of nine, with my NYU colleague Dr. Linda Moy becoming the first woman to lead the venerable journal *Radiology*).[2] Advanced imaging technology and know-how have only recently begun making their way into a number of heavily populated regions of the world. As critical as precedent, awareness of bias, and accessibility are to broadening the diversity of imaging research and development, though, another key ingredient is familiarity. As modern imaging methods, in all their dizzying variety, enter the public eye more and more, perhaps they can begin to attract the interest of a wider variety of young people. Perhaps imaging can finally shed the traditional trappings of an elite specialty and begin to remake itself in full living color.

All of this would allow imaging to come home to more people, but let's take one step further. Let's explore how imaging could come home in an even more immediate sense—that is, back to our own bodies. Artificial images are currently presented to our eyes just like any other natural scene, but aficionados of virtual reality, augmented reality, and mixed reality are already anticipating a time when manufactured imagery forms a new overlay to, or even a substitute for, our natural vision.

Current virtual, augmented, and mixed reality technology is just the beginning. The headsets offered by tech companies like Meta and Microsoft are getting progressively better at fooling our eyes—they can situate us in immersive gaming worlds and show us holograms that appear to hover over our real-life homes and workplaces—but they have barely scratched the surface when it comes to transforming vision. If we truly want to add seamlessly to natural vision, we may well need to replicate more of our eyes' and our brains' interactions with the real world, including jittery eye movements and various dynamic and/or multisensory cues. There is a lot of interesting research to be done to make the "metaverse" truly realistic.

Is aiming for visual realism the best we can do, though? Isn't it a little like using AI to copy radiologists—that is, an impressive and potentially useful feat, but not a particularly revolutionary one? Just as is the case for AI in imaging, if we move further upstream in the visual process, we can contemplate some more eye-popping innovations. Think about it: we are already walking around with heavy-duty tomography engines in our heads that are capable of synthesizing multiple disparate signals to form lifelike images. Our brains compare different views in our two eyes and combine them with a huge library of learned representations to intuit depth. Our visual processing centers juxtapose signals in different rod cells of our retinas to nail down fine shades of color. What if, instead of just piping new images into our eyes, we were to add some truly new signals into the mix?

CTRL-Labs, a brain–machine interface start-up acquired in 2019 by Facebook (now Meta) has developed noninvasive sensors that can pick up nerve signals traveling through the wrist and repurpose them to control computer displays, external devices or artificial limbs. If the reverse process could be mastered, then we could send new signals along existing sensory channels, or even directly into the brain, effectively hijacking biological inputs to create new artificial overlays. This is a challenging problem but one that scientists specializing in brain–machine interfaces are already poised to tackle. My imaging colleagues Kim Butts Pauly and Shy Shoham are part of a research community working to change the behavior of neurons in the brain from the outside using focused ultrasound. Other researchers are developing rudimentary implantable devices including artificial retinas and cochlear implants largely aimed at providing partial replacements for impaired senses of vision or hearing. Such interfaces could one day be fed not merely with coded visual or auditory signals but with ultrasound or x-ray signals. Our brains would have to learn the meaning of these signals, but once they did, we would be able to see through ordinary objects. Some distant variant of x-ray vision might one day be a superpower we all can share.

We've talked about changes that modern medical imaging has wrought upon our body image. Now consider if you will the changes in our sense of self that would accompany an artificially expanded sensorium. We would literally see ourselves anew. We could also expand our horizons in a strikingly direct way. Unlike our biological senses, which are hardwired through nerve channels that our bodies laboriously laid out as they grew, new sensory signal feeds would not all need to originate locally. They could be delivered wirelessly from widely distributed sensors positioned nowhere near our actual eyes. Rather than being restricted to our local perspective, we would see as far as our artificial eyes could see. Think about how your envelope of awareness tends to expand when you are driving a car or a truck: the boundaries of your "self" flex outward to include the edges of the vehicle, and you feel physically confined if you are caught in a tight space, even if your body has plenty of room to move. In the future, our sense of self could expand to encompass large swaths of physical space. We could shift our sense of scale at will, zooming in or out to the limits of our enhanced perception, or shifting the spectrum of what we see to suit our needs.

This notion of enriched sensory channels, along with a corresponding enrichment of one's sphere of awareness, has been explored extensively by writers of science fiction. In *Star Trek: The Next Generation*, the brilliant chief engineer Geordi La Forge (figure 12.1) is outfitted with a visor that translates a full spectrum of electromagnetic and other signals into his visual field. I know I was not alone in appreciating the notion that Lieutenant Commander La Forge, who was born blind, could leverage technology to transcend mere biological vision. Similar ideas fill the novels of Alastair Reynolds, whom I introduced in the last chapter. His books are populated by entire communities of people with richly amplified and widely distributed senses.

One such fictional community envisioned by Reynolds is a community of "Conjoiners": future humans who use networks of brain implants to communicate mind to mind. This may sound far-fetched to you, but from Mary Lou Jepsen's perspective, it is far from unthinkable. Jepsen is a celebrated inventor and former Big Tech executive who founded a start-up called Openwater in 2016. That same year, she visited our NYU imaging center for a workshop on the changing face of imaging. In her lecture on imaging systems of tomorrow, and in many subsequent talks in high-profile venues, she laid out Openwater's mission of using light not merely to image the brain but to map brain function in real time. One expressed goal of this work is to allow people to communicate directly mind to mind, without the bothersome intermediary of words. With virtual reality and its variants (technologies whose development she oversaw at Facebook), one can already see what someone else is seeing, but what Jepsen envisions is something else entirely—a sharing of perspective that extends beyond the merely visual. Skeptics legitimately question whether light alone is capable of such

**FIGURE 12.1** LeVar Burton as Lieutenant Commander Geordi La Forge in *Star Trek: The Next Generation*. La Forge's visor allowed him to see far beyond the visible spectrum.

*Source*: CBS Photo Archives via Getty Images.

heavy lifting.[3] Indeed, when I first heard Jepsen speak, my immediate reaction was, "That will never work." Getting light all the way through the head in a form useful for imaging is already a serious challenge—even without considering the level of neuroscientific understanding required to map a thought distinctively.[4] I would venture to predict that mind reading will not be on the menu tomorrow, or next year. After I had worked through my initial reservations regarding Jepsen's vision, though, my next thought was, "What would we do if it did work?" Whether you buy into Jepsen's plan or not, the gauntlet has once again been thrown. If one day we could convey thoughts directly from mind to mind, it would be a revolution in communication. If we could convey emotions or other states of mind, it would be a revolution in empathy. Such an opening up of the boundaries of self would constitute a technological transcendentalism worthy of Walt Whitman. "I am large," my future self proclaims. "I contain multitudes."

The bright future of seeing is full of possibilities: new equity and transparency, a rich technological metaverse, extrasensory perception, and shared states of mind. I've read far too much dystopian science fiction to take any of these possibilities at face value, though. So, here is the mirror image of how things might play out.

In one common sci-fi trope—call it the *Terminator* scenario—machine capabilities outpace our own, and machines take over the world, with disastrous consequences for their fleshy creators. Numerous people from various walks of life are already sounding the alarm on that front, including some leading lights in the AI firmament. Yielding up control of intercontinental ballistic missile batteries to neural networks certainly seems like an ill-considered idea to me. To be honest, though, I am not yet inclined to panic about the wholesale replacement of humans, for some of the reasons I've already laid out in previous chapters (missed predictions, shallow mimicry, and deficits in common sense to name a few). I take comfort once again from Yann LeCun, who places himself in direct opposition to AI "doomers."[5] Personally, I am less worried for now about humans being replaced than I am about humans being amplified, our darker natures enabled and disinhibited.

If we have learned anything from social media, it is that connecting people with technology doesn't simply promote transparency and enhance impartiality—it also cements prejudice and gives it room to grow. It would appear that artificial intelligences are not immune to such influences either. Artificial neural networks have already surprised people by learning all-too-human biases. An AI algorithm used by hospitals to predict which patients would likely need extra medical care favored white patients over Black patients. Another AI algorithm used in U.S. court systems predicted repeat offenses incorrectly twice as often for Black offenders as for white offenders. In 2015, it was shown that Amazon's automated hiring algorithm discriminated against women. In each case, the datasets and/or the metrics used to train the networks were incomplete, unfairly tilting the results toward those who had historically occupied positions of privilege. Similar biases can also make their way into imaging. For example, certain commercial facial recognition algorithms used by police in France, Australia, and the United States, among other countries, were recently shown to generate ten times more false matches in Black women than in white women or in men. Even when it comes to generating images, bias is an ever-present risk. Modern generative AI systems can easily depict offensive content with convincing photorealism. The ugly side of nominally free speech applies equally to visual forms of communication.

A partial democratization of imaging could also have mixed results when it comes to equity. Until and unless the cost and complexity of new imaging technologies are managed, personalized health monitoring could very well become a privilege that only those with sufficient money can buy. As a result, gaps in health equity could actually widen.

Meanwhile, the flip side of the new visibility afforded by imaging is an increasing inability to hide. While in some cases omnipresent personal cameras may deter abuses by those in power, the effect can be quite different when those in power own the cameras. In recent years, we have seen the rise of newly empowered surveillance states that use wall-mounted cameras, drones, or satellites to track the movements of ethnic minorities or hunt down the political opposition. The inevitable march of technology could make things very much worse indeed. Just as networks of satellites can currently monitor much of Earth's surface, clouds of connected sensors could all but eliminate private spaces. As was the case when evolving organisms first filled the early oceans, there will be eyes everywhere in our modern world, and many of those eyes could belong to predators. In some tech circles, the intriguing concept of "sensor dust" is bandied about. Some time in the future, a would-be spy could toss a handful of nearly invisible light-sensitive motes into the air, and those floating motes could put together a composite picture of whatever passes their way. For now, we can only imagine how such technology might operate. It doesn't take much to imagine its power, though—and power corrupts.

Moving back to medicine, the obvious downside to continuous medical monitoring is a new and unprecedented risk of unwanted medical surveillance. Let's say we succeed in building dynamic, individualized representations of health. Who, then, will own these representations? Who will be able to access them? It would be a truly existential threat to privacy if Big Brother and Big Tech were to gain access to your insides. If sensors in your clothes could be tied to your imaging history and used to predict your future, then why not sensors in your office chair? What would prevent the airport scanners you are required to walk through every time you board a plane from being used to probe your state of health? What would you do if your digital twin could spy on you? When I dream about the future of seeing, this is what gives me nightmares.

Tech companies already leverage the immense power of big data to track our movements and divine our desires as consumers. AI systems are already being trained to deduce a person's mood from photographs. It is not too great a stretch to imagine that optical sensors in a hat could be hijacked, perhaps correlated with functional MRI that tracks brain patterns, and used to interpret your state of mind. This is the dark side of Mary Lou Jepsen's telepathy project. If we can communicate with one another through our thoughts, then parties in possession of the right machines could also read our minds, whether with our consent or without it. Once again, this is not likely to happen tomorrow or next year, but with possibilities like this in the air it is important to think about appropriate protections well ahead of time.[6]

There are other risks which move in precisely the opposite direction to this unwanted transparency. New imaging technologies could also provide new opportunities to cover things up. Nowadays, you can go online and access powerful AI tools that allow you to fabricate convincing images of almost anything. The two images in figure 12.2 appeared together in a 2023 *New York Times* article titled "Can We No Longer Believe Anything We See?"[7] The article challenged readers to choose which of various pairs of images came from real sources and which were generated by AI. I won't make you guess. The iconic image on the left was taken in 1945 by the *Life* photographer William C. Shrout, documenting the celebrations that broke out when World War II came to an end. The image on the right was generated by the modern photographer and data scientist Victoriano Izquierdo by typing a prompt into Midjourney, one of a growing number of generative AI tools that creates images from text. One image documents historical events. The other shows people who never existed sharing a tender farewell kiss that never happened. Were you to see these images in a

**FIGURE 12.2** AI imitates life. Two images juxtaposed in Tiffany Hsu and Steven Lee Myers, "Can We No Longer Believe Anything We See?," *New York Times*, April 8, 2023. (*Left*) A 1945 photo by William C. Shrout depicting *Life* magazine photographer Alfred Eisenstaedt kissing a woman in New York's Times Square, emulating Eisenstaedt's iconic photo of a VJ Day kiss. (*Right*) An image generated by typing the following prompt into the generative AI program Midjourney: "Soldiers getting last kiss on ship before deployment to Egypt. Black and white. 1925. Men are on the boat and women are on the port. Old vintage photo."

*Sources*: (*Left*) LIFE Picture Collection/Shutterstock; (*right*) courtesy of Victoriano Izquierdo (https://victoriano.me/).

news report, would you be able to tell the difference? How would you feel, moreover, if one of the kissing figures had your own face? Increasingly convincing fake images and manufactured videos of celebrities are already making the rounds on social media, generating vehement reactions long before their artificial provenance is discovered. How easy it is for the most powerful imaging tools to become the perfect imaging weapons.

Figure 12.3 shows my own cursory attempt at using AI to manipulate photographic images. I had taken the picture on the left with an old mechanical camera decades ago—in the summer of 1982, to be precise—at Speaker's Corner in London's Hyde Park. Speaker's Corner, which has been a setting for public oratory since the 1800s, is a living symbol of free speech and open debate. As a child, I was struck by the rather modest ambitions of one utopian speaker (Utopia could be the answer to *some* of your problems?), so I snapped a photo of him standing on—I kid you not—an actual soapbox. Cut to 2022. I wanted to see how the photo would look with a more dystopian backdrop, so I visited OpenAI's DALL-E 2 site. DALL-E is

FIGURE 12.3 The future of seeing: bright future or dark mirror?

*Sources*: *(left)* Daniel K. Sodickson, 1982; *(right)* photo modified by the author using DALL-E 2 (OpenAI).

a visually focused cousin of OpenAI's ChatGPT, one of the generative chatbots employing large language models that have been turning decades-long norms of web search and millennia-long norms of writing, education, and communication on their heads of late. I uploaded my photo, highlighted a background area to fill in, and entered the following prompt: "dark, ominous apocalyptic landscape in the style of Hieronymus Bosch." In a scant few seconds, the image on the right appeared. The new background was not pieced together from existing images but was rather created de novo by the trained neural network that powers DALL-E, and was merged more or less seamlessly with the speaker from the actual photo.

DALL-E's generative effort in figure 12.3 was far more creative than what I would have managed on my own.[8] In this exercise, I was aiming for aesthetic effect more than convincing fakery, but, as figure 12.2 demonstrates, AI is proving increasingly capable of fakery as time goes on. Some hints of the algorithm may still be discerned if you look closely enough, but it is getting harder and harder to find chinks in the armor of believability. Tools for image manipulation are even being billed as selling points for new devices: some smartphone manufacturers, for example, advertise camera apps that can edit people out of photos seamlessly with the swipe of a finger.

This trend may end up affecting more than smartphone sales. I'm talking about a potentially existential threat to truth. Photographs were once considered documentary evidence, but in the era of deep fakes, how—as *New York Times* reporters and many others have already wondered—can we continue to trust the evidence of our eyes? To make matters worse, if humans do one day manage to engineer direct overlays to our biological vision, then our eyes themselves could eventually be hacked. We could be presented with doctored visual data streams, and, if it were done well, we might never know the difference. Who needs fancy stealth technology, which absorbs light or bends it around hidden objects, if you can directly control what people see? If seeing is believing, then whoever controls the content of vision will control people's hearts and minds.

When it comes to distortions of the truth, one doesn't even need a bad actor controlling what people see. All one needs is to allow people to see only what they want to see. The precedent of social media once again applies. When news feeds are individually tailored, and when users freely choose to follow those they like, consensus reaches only as far as each friend group. Like hews to like, and dogma is amplified rather than debunked. The same can be expected to hold true for seeing. In some of his stories, Alastair Reynolds imagines what things would look like if people could choose what they want to see. In these speculative futures, individuals occupy customized levels of "abstraction," with neural interfaces or implants feeding them carefully tailored sensory data. The "real world" is filtered or obliterated as each person sees fit. You can view this state of affairs as a triumph of autonomy, identity, and self-expression—or you can see it as an abject failure of empathy. In this dark mirror of the future, we enter a new kind of technological solipsism in which each of us floats alone at the center of our own fabricated universe.

If we were to take the technology of seeing to these dark places, I worry that what we see would be *all* that we get. I worry that we would all get precisely what we ask for—and that this would be the worst thing that could happen to us.

I know. That's a bit much to digest. Here I am telling you that imaging is something to know and love and embrace without fear, and then I drop two extreme futures on you and leave it to you to choose. You are well within your rights to ask, "What gives, Mr. Imager?"

So ask me, then. Which *will* it be: bright future or dark mirror? I cannot tell you that for sure, but I can tell you where I stand. I stand with the light.

You've spent long enough with me to know that I am an optimist. Let me now tell you why. Any personal temperamental disposition aside, I think it is because I am a medical imager, and the example of medical imaging gives me hope. Those of us who make our vocation out of medical imaging know that our images are the result of artifice. We know that the images guiding doctors and patients are not given but made. Nevertheless, we take pains to ground them in truth because in medicine truth matters. The right images, at the right time, can make a real difference.

Every technology has the capacity to be used for good or for ill, and imaging is no exception. On balance, though, medical imaging has done far more good than harm. I challenge anyone to convince me otherwise. Of course there are missed diagnoses, bungled examinations, unjustified screening tests, and bad outcomes. The news from some scans may be difficult to hear, and the costs may be difficult for some medical systems to bear. But talk to anyone whose life was saved by imaging. Tally the families and friends, in their millions, whose loves have been preserved. Who would want to go back to the dark ages before x-rays and tomography? Who would not wish to share the benefits of modern medical imaging modalities with those who are currently denied them? A future of widely accessible medical imaging and cooperative early warning could also do a great deal more good than harm—if we can find our way to that future.

I am also optimistic because I am a scientist, and I believe that knowledge matters. Astronomy, microscopy, photography, tomography—all of these have opened our eyes to more things in heaven and Earth than ever we might have dreamt of in our philosophies. Imaging advances knowledge. Knowledge, as Sir Francis Bacon famously opined, is power. And power, of any kind, can be used for good or ill. To quote Tom Stoppard, "Information is light. Information, in itself, about anything, is light."[9] The question, then, is where we shine that light.

This is what we need to choose—you and I, and everyone together. How will we use the power that new forms of imaging will bring us? Will we take pains to ground new ways of seeing in truth, or will we let them stray? Will we stay in one comfortable perspective to the exclusion of all others, or will we open ourselves to others? Seeing occupies a slippery interface between technology, psychology, and culture that can be challenging to manage—but knowledge is power. So, keep your eyes open.

Our machines will not dictate the path to utopia or dystopia. Right now, the choice is ours. I believe in the bright future of seeing. We might just need to work together to get there.

# Epilogue

## *The Continuing Story of Imaging*

The kiss of his memory made pictures of love and light against
the wall. Here was peace. She pulled in her horizon like a great
fish-net. Pulled it from around the waist of the world and draped
it over her shoulder. So much of life in its meshes! She called in
her soul to come and see.

—ZORA NEALE HURSTON, *THEIR EYES WERE WATCHING GOD*

A uthors, artists, and scientists sometimes have their differences, but on this they all agree: so much of life is a matter of perspective.[1] And that is why, as I see it, the story of imaging is a story worth telling. To say that imaging changes your perspective is almost a matter of definition. Imaging *is* perspective. The story of imaging, meanwhile, is not just a story of experts and their tools. It is a broader human story. I hope this book has helped to bring that story home to you.

You and I began our travels together with a whirlwind tour of the history of seeing. Seeing, which is to say spatially oriented sensing, serves a deep-seated need baked into our biology and as primal as the need for food and shelter. We make pictures of our world in order to understand it, and to find our place in it. We seek, therefore we see. Information is light.

It is really no surprise, then, that from the beginning, humans have restlessly sought out new ways of seeing.[2] Science, at its root, is the art of seeing clearly. Art, meanwhile, can be defined as the science of seeing differently. As both art and science advanced, people learned new ways to record what they beheld in the form of images. They learned to share perspectives, to copy vision, preserve it, and enhance it. That is the history of imaging.

When it comes to furthering understanding, artificial imaging picks up where natural vision leaves off. Thanks to a remarkable progression of technologies and transformations, our understanding is no longer limited

227

to our immediate surroundings—it extends both farther out and deeper inside. The magnification that resulted from bending light gave us astronomy and microscopy. Cameras gave us a faithful record, captured first on film, then in the abstract form of digital data. X-ray projections allowed us to see the insides of intact bodies. Tomography stitched disparate projections together, laying out the body's secrets in exquisite cross section.

As you have witnessed, imaging has a long track record of disruption. Telescopes cemented the Copernican Revolution. The view through microscopes led inexorably to the world of the microbe, the molecule, and the atom. Observations of inner and outer space are today encapsulated in fundamental theories connecting the indescribably small to the unimaginably big. At the same time, imaging has catalyzed fundamental changes in the way we live our lives. Our health is probed with pictures. We communicate via video. Each week, the number of images shared online is roughly three times the human population of our planet.

The hits will keep on coming, because the story of imaging is far from over. It is being written and rewritten day by day, by scientists and clinicians, by entrepreneurs and patients, by me, and also by you. In addition to being a story worth telling, then, the story of imaging is an important story to understand. In addition to helping us sort out how we got where we are, it can help us to see our future.

This is not to say that the future I have imagined in the second part of the book is sure to come to pass. History teaches us as much. In 1909 and 1910, the Austrian journalist Arthur Brehmer assembled a collection of predictions by prominent experts about what the world would look like in one hundred years.[3] Considered from the vantage point of the world we know today, a little more than a century later, these predictions are both quaint and illuminating. They include some notable successes, such as an uncanny forecast of the impact of wireless technology. They also include some equally notable failures, including substantial underestimations of our capacity for high-speed transportation, and a triumphal expectation that, in our time, humankind would be free of disease. In my view, Brehmer's collection illustrates a general tendency to underestimate shifts in technology while overestimating advances in medicine. With health, it is comparatively easy to predict vast leaps, since we can easily envision what a revolution in health would look like: diseases we know all too well would simply disappear. The biological systems that affect health, however, are complex and built for homeostasis. By design, they resist sudden changes.

With technology, on the other hand, the sky's the limit, and new technologies can often develop in directions we simply didn't envision.

When it comes to the future of imaging, some things are clear. So long as there is support for science, discovery will continue, as it has throughout history. New imaging modalities will be spawned every time a new probe comes to light, and once-new imaging modalities will be relentlessly refined, providing ever-clearer windows onto the world. Even for existing imaging modalities, we can fully expect to see new information content in our pixels. Imaging history has certainly taught us that new and surprising forms of image contrast are there for the taking, if one is just bold enough to reach for them, and just stubborn enough to apply the necessary software, hardware, or wetware.

All of this is business as usual for imaging. Recent trends, though, suggest that something more unusual may also be brewing—something strange but at the same time oddly familiar, harking back to our longer history of seeing. We may finally have the tools to emulate our senses, and our brains. We may finally have the tools to change them. As a result, our imaging devices will be smaller, more connected, more integrated into our lives, and potentially more disruptive to them. New transformations may take us all the way from bending light to mapping states of mind, with dramatic implications for our politics, our communications, our relationships, and our identities. The next imaging revolution may look something like a new stage of evolution.

Imaging has been with us from the beginning, and it will be with us far into our future. It does not live merely in the arcane ministrations of astronomers or in the fearsome prognostications of medical doctors. It lives in us. It is deeply interwoven into our history, and it is inextricable from our senses of ourselves. Imaging is a little bit like breathing. It is second nature to us, even if we generally do not pay it much attention. Like it or not, for better and for worse, we are all creatures of imaging. This is our story.

# Acknowledgments

Inscribed in the book of my memory, and also on the inside of my wedding band, are the three Latin words that serve as a starting point for Dante's *La Vita Nuova*. *Incipit vita nova*: thus begins a new life. In the last two years or so, I have come to understand that the origin of a book can be something like the origin of a life.

The story of imaging that I have sought to tell you is a story that has been growing on me for a long time, and so the origins of this book have a lot in common with my own origins. My first thanks, therefore, must go to my parents, Lester and Isabel Sodickson. They are my authors, and my very first readers. Theirs was not a singular act of creation but a sustained light of attention, conversation, partnership, and understanding that none of time's gradual wear or sudden tragedies could dim. That light sustains me still. Mom is a social worker, Dad is a physicist, and together they make the most remarkable team. From the earliest days of my youth, I could see my father's eyes light up whenever he was talking about how things worked. They still do to this day. And anyone can see the gleam in my mother's eye when she talks about what makes people tick. My parents' lessons in curiosity and caring have opened up worlds to me.

Mom and Dad continue to be in my innermost circle of readers, and their complementary perspectives on drafts of this book have been invaluable—inspiring me, for example, to lead with stereo vision in chapters 4 and 10, and helping me to chart a course between bright future and dark mirror in chapter 12. In elementary school, I once ended a short story I had written with the words "On the table was a finished weaving." Here you go, then, Mom and Dad. The story is far from over, but this little bit of weaving is finally done.

In the volume of memory that begins at the turn of the millennium, one person looms large over all others. I met Sarah Lieberman in March of 2001, and by April of 2002 her wedding band made a matched pair with my own. Sarah sometimes begins a book by flipping directly to the acknowledgments section. This says so much about her. Sarah, also a social worker by vocation, is interested not only in the end product on the page, but also in the backstory of how it came to be. The same is true, and even more so, for people. Sarah is genuinely fascinated by people's stories, and she is a past master at eliciting them. She draws people in with her warmth, and she draws them out with her attentive questions. I was recently sitting with Sarah and our daughter, Hannah, on a Metro North train to Manhattan when Hannah nudged me and tilted her head toward Sarah: "Mom is making friends again." Indeed, Sarah was engaged in lively conversation with a family from Texas, who were busy pouring out their tales to her. When I meet someone new, or even when I am talking turkey with a colleague, I have trained myself to slow down for a moment and think, "What would Sarah ask?"

It is said that there is no revelation without agitation, and, when it comes to this book or to our life together, Sarah has lived with my every agitation. She has tolerated a shower wall covered with notes scrawled in waterproof pencil. This is partly her own fault: she bought me my first pack of AquaNotes™, which have become some of my most indispensable writing tools. (Writers everywhere, take note: your scribblings need not end when your daily ablutions begin.) As readers go, Sarah has been my alpha and my omega. She reined me in when I got too flighty or too technical, and she encouraged me when perspectives resonated with her. Sarah—this acknowledgment is all yours. You and I have navigated so many new beginnings together. It is my privilege to share this one with you.

Glenn Magid is the next reader to whom I owe special acknowledgment. I first met Glenn more than three decades ago during my graduate studies at Harvard and MIT. Glenn's graduate program in ancient Sumerian language and civilization looked almost nothing like my own, but we hit it off instantly, becoming lifelong friends and conversation partners. *Magid* is derived from a Hebrew term for a mystical storyteller, and, fittingly, Glenn is an inveterate turner of phrases, whose conversational gymnastics have long meshed with my own. He is an even more inveterate page-turner, who can never be found without a book. A more voracious reader I defy you to produce. Though not formally trained in engineering, Glenn has built a unique apparatus that props up his latest book together with a cup of coffee so that he can read freely while walking. If you visit the greater Cambridge area, you may encounter him read-walking his regular beat. He has been an alpha reader of mine from some of the earliest drafts onward. In many ways, this book was written with Glenn in mind.

I am grateful to a long list of other friends and colleagues for contributing to the genesis of this book. I thank the remarkable community of the International Society for Magnetic Resonance in Medicine for being my intellectual home away from home and my professional extended family for twenty-eight years now. I credit my time in ISMRM leadership with focusing me on the importance of telling the story of imaging to stakeholders far and wide. Early in my term as ISMRM president, I was invited to deliver the New Horizons Lecture at the RSNA's annual meeting in late November of 2017. I thank Dick Ehman—a respected colleague, a former ISMRM president, and that year's RSNA president—for entrusting this lecture to me. The storyline I conceived in the months leading up to the lecture would serve as a concrete seed for this book.

Once it had been planted, that seed grew gradually until it left room for little else. This left me with the challenge of figuring out how to bring a book to life. György Buzsáki, my NYU colleague who has authored brilliant popular science books on the workings of the brain, served as a role model and provided helpful guidance. Likewise for Elkhonon Goldberg, another noted explicator of the human mind and brain, who introduced me to Michelle Tessler, who in turn took a chance on a new author and became my agent. Imaging is not a topic that automatically sells itself, so Michelle and her intern, Hannah Wiles, helped me to construct a book proposal with a chance of getting noticed. Michelle then connected me to Miranda Martin, who noticed. As my editor at Columbia University Press, Miranda has served as a champion of and shepherd for my story ever since. I will forever be grateful to Miranda and Michelle for taking my sparse projections of what this book could be and envisioning it in depth, trusting a first-time author to fill in what was missing.

Various people have made concrete contributions to what I ended up filling in. First, there are those who allowed me to include woefully cursory profiles of them in chapters 6 and 7. My thanks to Urvashi, Roberta, Mary, Bob, Rod, Jürgen, Walter, Michael, and Karla for their willingness to live with what I have fashioned in their image. Thanks also to the individual imagers and visual artists who gave me permission to share with you the fruits of their creative labors as part of this book's illustrations: Rizwan Ahmad, Franz Anthony, Greg Chang, Martijn Cloos, Dorin Comaniciu, Clarissa Cooley, Yulin Ge, Veronica Falconieri Hays, Joe Helpern, Victoriano Izquierdo, Dan Ma, Graham Murdoch, Pawel Slabiak, Pippa Storey, Andrew Webb, Graham Wiggins, and Joseph S. K. Woo. Shy Shoham went out on a limb to start NYU's Tech4Health Institute with me, and he also lent me the monograph *Animal Eyes*, which helped to inform parts of chapter 1. Pratik Mukherjee, a college roommate of mine who trained in neuroscience and radiology and who has gone on to become a thought leader in brain imaging, is the one who turned me on to the visual system of horseshoe crabs (see chapter 10), and he has been feeding me with fascinating new bits of information related to vision ever since he learned I was writing this book. Hersh Chandarana served as my right-hand man in radiology research leadership at NYU for many years and, in addition to being another reader of early drafts, he has been my principal partner in envisioning the everywhere scanners described in chapter 11. It took Yann LeCun all of five minutes to convince me that self-supervised learning was the tool I had been looking for to connect the dots of everywhere scanners, and I thank him for the time he has made to talk. My conversations with him, not to mention his many articles and lectures, have informed various parts of chapters 10 and 11. Florian Knoll, a colleague and former postdoctoral fellow of mine who helped to launch the modern era of AI-accelerated imaging, made me aware of the early-twentieth-century Austrian futurist survey mentioned in the epilogue. I thank Rania Assas, my assistant and my friend over decades filled with joyous births, devastating deaths, superstorms, and other sea changes, for helping me to create, preserve, and protect blocks of time for writing. I am grateful to the editorial and production teams at Columbia University Press, including Miriah Ralston and Kathryn Jorge, as well as to Ben Kolstad and the KnowledgeWorks Global team, for all they have done to transmute my words and images into the form of a book.

Then there are those who have contributed to my own understanding of imaging. I have mentioned some of them in passing, whether by name or in larger groupings. You have met, if briefly, some of the lively cast of characters who congregated at Beth Israel Deaconess Medical Center in the late 1990s and early 2000s, and who gave me my first glimpses of the wide horizons of imaging. To the research group that eventually gathered around me in Boston, I owe countless lessons in leadership, as well as endless conversations about signal-to-noise ratio. At NYU beginning in 2006, I found a remarkable family of colleagues and friends far too varied to characterize in a pithy phrase or two, and far too numerous to name. You know who you are, and you know what you mean to me. I will do my best to find other ways to tell you.

I thank the National Institutes of Health, the National Science Foundation, and their peer organizations around the world—along with you, the global taxpayer—for supporting the work of all these imaging colleagues, and many more, over the course of these long years. Thanks to you, the circle of imaging continues to expand.

Zooming out just a little bit more, there are a few people I must thank for the life lessons they have imparted and the companionship they have provided. Person 1: During my high school years at the Roxbury Latin School in West Roxbury, Massachusetts, I encountered the force of nature known as Mo Randall. In ninth grade English class, Mr. Randall schooled me mercilessly in constructing the perfect paragraph. He was my guide to Virgil in eleventh grade Latin, leading the class line by line through hellish syntax and heavenly verse. Then in senior English, Mo let me write whatever I felt like writing, flouting all the rules I could think of. I thank Mo for setting me on the path of principled storytelling. Responsibility for the glaring imperfection of my paragraphs is entirely my own. Person 2: After graduating from Roxbury Latin, I found my way to Yale, where I met a friend whose constancy would see me through all the many years to come. Dan Sidney is inquisitive, considerate, quirky, and kind. We roomed together in our sophomore and junior years, and we have been fellow travelers ever since, in life as well as in imaging. Persons 3 and 4: Apart from Mom and Dad, my most abiding travel partners have been my sister, Deborah, and my brother, Aaron. From the home we shared to the new families we formed, they have been my constant companions in learning to see the world in new ways. Aaron is also an imaging compatriot and another early reader.

*Incipit vita nova.* Hannah arrived in my life in 2004, and Noah followed in 2006. Sarah and I have seen Hannah's and Noah's stories unfold from the very beginning. The last word here must go to them. Hannah: my bold, principled, change-resistant yet status-quo-hating, underdog-loving and difference-embracing, scenery-chewing, puzzle-solving, computer-brained, big-hearted daughter. And Noah: my enthusiast par excellence, my insatiable student of science and history, my coder of codes and framer of arguments and connoisseur of films and fine cuisines, my wide-eyed visionary son. Hannah and Noah, you see things as I can't. You are growing up in a world transformed by miraculous technologies, filling your senses with staccato signals of constant information. You walk the line between dark mirror and bright future, and, if anyone can, you will steer us toward the light, devising wondrous new technologies and transformations, or turning such tools to worthy tasks, and imagining how they can be used to see things differently. Hannah and Noah: the future of seeing belongs to you.

# Notes

## 1. THE NATURE OF SEEING

1. A brief note on language: throughout this book, I refer to how "your eyes" or "our eyes" work. I realize that this language, though intended to personalize the narrative, is not inclusive. People's eyes work to varying degrees. However, even if you do not rely principally on your biological vision to navigate the world, you still have a stake in the story of seeing. I would even argue that you may be one of its pioneers. After all, though artificial imaging may have taken cues from biology, it is designed to pick up where ordinary eyes leave off. As we will explore in part II, the future is sure to transcend what today's eyes can see.

2. Admittedly, at the limits of near or far vision, some eyes may require the aid of corrective lenses using principles we will explore further in chapter 2.

3. Ed Yong, *An Immense World: How Animal Senses Reveal the Hidden Realms Around Us* (Random House, 2022); Ed Yong, "Inside the Eye: Nature's Most Exquisite Creation," *National Geographic*, January 5, 2016, https://www .nationalgeographic.com/magazine/article/evolution-of-eyes.

4. Charles Darwin, *On the Origin of Species* (Cassell, 1909; originally published in 1859 by John Murray). This quote and the one in the next paragraph were taken from Darwin's chapter 6, "Difficulties on Theory," in the subsection entitled "Organs of Extreme Perfection and Complication." Interested readers may find the passage in question and related arguments on page 163 of the free 2019 Amazon Classics edition or on page 172 of the 1909 Cassell edition reproduced in Google Books.

5. Michael F. Land and Dan-Eric Nilsson, *Animal Eyes* (Oxford University Press, 2002).

6. Dan-Eric Nilsson and Susanne Pelger, "A Pessimistic Estimate of the Time Required for an Eye to Evolve," *Proceedings of the Royal Society B* 256, no. 1345 (1994): 53–58.

7. To make a light-sensitive patch, just take a few molecules, cooked up in the primordial soup, that can change their configuration when exposed to light, and fix them in place.

8. The story of what pugnacious mantis shrimps do with all those chromophores is a captivating mix of colorful speculation and murky understanding. See Yong, *An Immense World*, chapter 3.
9. They say that Gossip can circle the globe before Truth gets out of bed. By this measure, light is an enviably swift gossip.

## 2. AUGMENTING NATURE

1. As shown in the figure, the image projected onto the back wall of a camera obscura will be upside down and backward. This is because light rays must converge to a point at the location of the pinhole in order to be admitted into the room, and those rays necessarily cross at the point of convergence. Thus, the resulting image is inverted by the time it reaches the far wall.
2. The text of Plato's *Republic* actually describes shadows produced by objects interposed between a blazing fire and a cave wall. However, the text also envision a cavern mouth open to the light above. If Plato had taken away the firelight and let reflected light from objects outside the cave do the work, he would have had a bona fide camera obscura, and a hundred generations of students might have thought of "light pictures" rather than "shadow pictures" when pondering the nature of reality.
3. Note that in so-called ray-tracing diagrams, such as those in figure 2.2, the direction of the arrowheads indicates the direction in which light waves are traveling.
4. At any other distance, light coming from each point on an object either will have failed to converge completely or will have begun to spread out again, and parts of the image will blur together.
5. On the controversy surrounding the invention of the telescope, see Albert Van Helden, "The Invention of the Telescope," *Transactions of the American Philosophical Society* 67, no. 4 (1977): 1–67. As it turns out, all these squabbling humans were actually quite late to the game. In chapter 2 of *An Immense World*, Ed Yong points out that jumping spiders had evolved a two-lens magnifying configuration in their eyes millions of years earlier.
6. For distant objects, the primary image shows up at the focus of the objective lens.
7. Alan Chapman, *England's Leonardo: Robert Hooke and the Seventeenth-Century Scientific Revolution* (CRC Press, 2004).
8. Freeman Dyson, "The Scientist as Rebel," *New York Review of Books*, May 25, 1995.
9. Ironically enough, for this first real-life image Niépce chose to imitate art, exposing his bitumen-coated surface to an illuminated engraving of Pope Pius VII. In other words, rather than copying the image of a natural scene faithfully, he created a faithful heliographic copy . . . of a hand-carved image!

10. Roger Watson and Helen Rappaport, *Capturing the Light: The Birth of Photography, a True Story of Genius and Rivalry* (St. Martin's Press, 2013).

## 3. SEEING IT THROUGH

1. Bettyann Holtzmann Kevles, *Naked to the Bone: Medical Imaging in the Twentieth Century* (Addison-Wesley, 1997), 23. I have relied heavily on Kevles's account of the discovery and rapid adoption of x-rays. She also traces the subsequent history and societal impact of tomographic imaging modalities, and her thoughtful work has informed some of the historical backdrop for chapter 4.
2. "The Nobel Prize in Physics 1901," NobelPrize.org, accessed March 26, 2025, https://www.nobelprize.org/prizes/physics/1901/summary.
3. Kevles, *Naked to the Bone*, 23.
4. Kevles, *Naked to the Bone*, 35. The hand image in figure 3.3 is also reproduced in *Naked to the Bone*, where Kevles recounts the story of this accident and the ensuing imaging examination.
5. Indeed, the first report of a fractured wrist imaged using x-rays was published in the journal *Science* in February 1896—as part of the very same issue in which Röntgen's original report, "On a New Kind of Rays," appeared. Peter K. Spiegel, "The First Clinical X-ray Made in America—100 Years," *American Journal of Roentgenology* 164 (1995): 241–43.
6. According to Kevles, "skin became just another wrapping, something to be removed to reach what was more valid beneath it." Kevles, *Naked to the Bone*, 28. See also chapter 6 of *Naked to the Bone*, "X-rays in the Imagination."

## 4. SLICING WITHOUT CUTTING

1. There is no known relation between Johann Radon's family name and the radioactive gas known as radon, discovered in 1899 and experimented on by Marie and Pierre Curie, among others.
2. Note that, for visual clarity, I have chosen to show regions of high x-ray attenuation as bright rather than dark, a convention derived from x-ray film that is maintained in CT images.
3. If you're a stickler, you might have noticed that the circles at the bottom of figure 4.1 are a little blurry. Good eye! As it happens, pure back projection doesn't quite yield an exact solution to the projection problem because of all the fuzzed-out intensities in the background of the image. In practice, it takes only a small tweak—a so-called filtering step—to render back projection nearly exact. So, the conceptual picture in figure 4.1 remains a good base for understanding.
4. Bettyann Holtzmann Kevles, *Naked to the Bone: Medical Imaging in the Twentieth Century* (Addison-Wesley, 1997), 108.

5. In a 1992 article, Cormack stated, "It was only by chance that in 1970 I became aware that this problem had been considered and solved by Johan Radon." Allan M. Cormack, "75 Years of Radon Transform," *Journal of Computer Assisted Tomography* 16, no. 5 (1992): 673.

6. Allan M. Cormack, "Early Two-Dimensional Reconstruction and Recent Topics Stemming from It," Nobel lecture, December 8, 1979.

7. Hounsfield set his computer to the task of solving large sets of simultaneous equations to determine how many x-rays had been blocked at each position in the imaged object. These operations were a little different from the procedures derived by Cormack and Radon and from the back-projection process illustrated in figure 4.1, but they accomplished the same thing.

8. Hounsfield's first CT article wasn't published until two years after the first clinical CT scan had been performed. Godfrey N. Hounsfield, "Computerised Transverse Axial Scanning (Tomography) I. Description of System," *British Journal of Radiology* 46 (1973): 1016–22.

9. Hounsfield noted this wide dissemination in his Nobel lecture. Godfrey N. Hounsfield, "Computed Medical Imaging," Nobel lecture, December 8, 1979.

10. As Kevles explores in chapter 7 of *Naked to the Bone*, the Australian astronomer Ronald Bracewell in the 1950s, the American neurologist William Oldendorf in the early 1960s, and the American radiologist David Kuhl a few years later each made notable strides in mathematics, hardware, and even the generation of human images.

11. In actuality, Lauterbur used an iterative variant of back projection that worked well with a limited number of projections, rather than the simple back-projection procedure illustrated in figure 4.1.

12. Lauterbur's classic illustration was actually my inspiration for figure 4.1.

13. Actually, quite a bit of work went into this slender publication (Paul C. Lauterbur, "Image Formation by Induced Local Interactions: Examples Employing Magnetic Resonance," *Nature* 242 (1973): 190–91.) Lauterbur would later share that his original submission to *Nature* was "summarily rejected." He had to submit a revision, discussing potential medical applications, before the journal took an interest in the paper. "Almost thirty years later," Lauterbur observed wryly, "*Nature* publicly celebrated its appearance." Paul C. Lauterbur, "All Science Is Interdisciplinary—From Magnetic Moments to Molecules to Men," Nobel lecture, December 8, 2003, 248. Students everywhere, take note. Sometimes, even the best ideas are not immediately recognized for their underlying brilliance. Sometimes, persistence is called for—and rewarded.

14. Being a solid-state physicist, Mansfield took inspiration from an established tool of physicists known as x-ray crystallography, which can determine the microscopic structure of crystalline solids. Peter Mansfield and Peter K. Grannell, "NMR 'Diffraction' in Solids?," *Journal of Physics C: Solid State Physics* 6 (1973): L422–26.

15. Allen N. Garroway et al., "Image Formation in NMR by a Selective Irradiation Process." *Journal of Physics C: Solid State Physics* 7 (1974): L457.

16. X-rays can liberate electrons from their orbits around atomic nuclei, leaving behind positively charged ions. The deleterious effects of ionizing radiation on living tissue were not realized at first, but it did not take long for reports to circulate describing strange burns associated with intense x-ray exposure. In the longer term, as it would come to be discovered, high doses of ionizing radiation are associated with an increased incidence of certain cancers. Marie Curie certainly paid the price for her work with ionizing radiation: she died of radiation-induced leukemia. The question of acceptable radiation doses from medical imaging techniques using ionizing radiation has been a subject of legitimate attention in the field and in the press at various times, and it has also engendered a fair amount of fear among laypeople. As a patient, it is always good to be aware of the safety profile of the probes used for imaging and to ask any questions you might have about radiation controls. But it is also important not to let a general fear of radiation keep you from potentially life-saving scans.

17. William A. Edelstein et al., "Spin Warp NMR Imaging and Applications to Human Whole-Body Imaging," *Physics in Medicine and Biology* 25, no. 4 (1980): 751–56. As early as 1975, the future chemistry Nobel laureate Richard Ernst and colleagues had dreamed up a way of rastering through frequencies rather than angles to generate magnetic resonance images. Anil Kumar et al., "NMR Fourier Zeugmatography," *Journal of Magnetic Resonance* 18 (1975): 69–83.

18. The Nobel Prize for work on MRI was awarded in medicine, but by that time developments in magnetic resonance had already garnered Nobel Prizes in physics and chemistry. The 2003 prize brought the total haul for magnetic resonance to five Nobels, awarded to seven people spanning diverse areas of physics, chemistry, biology, and medicine.

19. It was, in fact, Damadian's team that published the first-ever magnetic resonance image of a live human body in 1977.

20. Chapter 9 of Kevles's *Naked to the Bone* provides an in-depth recounting of the history of PET development.

21. Hevesy received the Nobel Prize in Chemistry in 1943 for his pioneering use of radioisotopes in biological science.

22. As our story progresses, you will learn that MRI eventually acquired a robust capability for functional imaging, but that capability was developed over the course of decades, whereas PET was born with it.

23. Other tomographic imaging modalities in addition to PET are now used for molecular imaging, generally through the use of tailor-made "contrast agents" that introduce chemical and molecular specificity into images. Once again, though, PET had its origin in contrast agents, in the form of radiotracers.

24. A quick caveat: echolocation actually does come naturally to some humans. In *An Immense World*, Ed Yong tells the story of Daniel Kish, a survivor of

childhood eye cancer who uses echoes of vocal clicks to observe the world, and who teaches others in the blind community how to echolocate. Ed Yong, *An Immense World: How Animal Senses Reveal the Hidden Realms Around Us* (Random House, 2022), chapter 9.

25. Radar, also developed during World War II, is another range-finding technique that uses principles of echolocation learned from other species. In place of sound waves, though, radar employs radio waves.

26. See, for example, D. N. White, "Neurosonology Pioneers," *Ultrasound in Medicine and Biology* 14, no. 7 (1988): 541–61; James Zagzebski, "History of Ultrasound Imaging," paper presented at the 54th Annual Meeting of the American Association for Physics in Medicine, Charlotte, NC, August 1, 2012, https://www.aapm.org/education/vl/vl.asp?id=397, accessed March 30, 2025; Kevles, *Naked to the Bone*, chapter 10.

27. This sensitivity to changes in density explains why ultrasound probes are generally placed directly on the skin, often following application of a gel that approximates typical tissue densities. Without this direct contact, too much of the ultrasound signal would bounce back from the sharp interface between low-density air and high-density tissue, and not enough would be left to resolve interior structures clearly. The density differences between air and tissue actually result in some fascinating differences between the echolocation capabilities of bats and dolphins, as Yong explains in chapter 9 of *An Immense World*. Since bats operate in air, reflections from higher-density external surfaces make up most of what they can detect with echolocation. Dolphins and whales, on the other hand, are sea creatures, and the density of seawater is a close match to that of tissue. This means that the vocalizations of dolphins and whales can penetrate tissue surfaces effectively, allowing the animals to "see" inside other sea creatures, picking out low-density air cavities and high-density bones with the sonic equivalent of x-ray vision. Yong calls these aquatic echolocators "living medical scanners."

28. The problem of deducing what must have produced a signal based on knowledge of how that signal was generated falls under the general category of mathematical problems called inverse problems.

29. Victor R. Fuchs and Harold C. Sox Jr., "Physicians' Views of the Relative Importance of Thirty Medical Innovations," *Health Affairs* 20, no. 5 (2001): 30–42.

30. Mammography (breast imaging with x-rays) and ultrasound, by the way, showed up at number five and number eleven on the list, respectively.

## 5. WHAT'S IN AN IMAGE?

1. One could just as well use hexagons, or nesting figures from an Escher print, or just about any set of shapes of any size that fit together seamlessly. Remember the jaggedly blocky butterfly's-eye view way back in chapter 1.

2. When pixels are arranged in a regular fashion, as in figure 5.2 and in most synthetically generated images, it is quite easy to connect quantities such

as resolution and field of view. The nominal spatial resolution in any given direction is just the field of view in that direction divided by the number of regular pixels arrayed along the same direction. Let's say the view of the brain in figure 5.2 is 200 millimeters high and 150 millimeters wide (field of view: 200 mm × 150 mm). If the image contains 200 × 150 regular pixels, then the nominal spatial resolution is 1 mm × 1 mm. Though such simple intuitions can be helpful in parsing the structure of images, the professor in me is moved to caution you not to mistake nominal resolution for actual resolution. I drill it into my students that actual spatial resolution depends on both the physics of image collection and the mathematics of image reconstruction.

3. A brief note for fans of physics and philosophy: My use of the term *spacetime* here is of course a nod to Einstein. That said, we are free to move back and forth as we wish through the spacetime represented by images, free of the relativistic constraints that governed actual events as they happened. The notion of slicing through space and time at will, meanwhile, recalls Immanuel Kant's understanding of space and time not as entities in themselves but rather as indices of experience. Indeed, the notion of images as representations of experience in an ordered spatial and temporal grid can be seen as Kantian in a remarkably literal way.

4. The discovery of cosmic microwave background radiation is a famous case in which one person's noise became another person's signal. This pervasive radiation was first detected as noise degrading the routine signal from radio telescopes until, in a Nobel-worthy turn, Robert Woodrow Wilson and Arno Allan Penzias recognized it as quiet reverberations from the big bang. Images of cosmic background radiation now provide us with rough spatial maps of our universe near the very beginning of time.

5. "Crab Nebula," NASA, accessed April 22, 2023, https://www.nasa.gov/multimedia/imagegallery/image_feature_567.html.

6. Sir Frederick William Herschel, an astronomer and musician who built telescopes and cataloged planets and nebulae with his sister Caroline, is credited with the discovery of infrared light in 1800. In an unexpected observation that had many of the hallmarks of Röntgen's later discovery of x-rays, he noticed that the heat generated in a thermometer by sunlight that had been spread out by a prism extended beyond the red end of the visible spectrum. He posited the existence of invisible heat rays, which he soon found could be bent, reflected, and absorbed in the same ways as visible light. And, lo and behold, there was a new probe that would one day be used to make images.

7. Early astronomers from ancient Greece onward are known to have favored a concept called the "harmony of the spheres," in which the motions of celestial bodies were thought of as a kind of music. In a way, multicolor images containing information captured at multiple frequencies can be thought of as a visual analogue to this music. Some modern astronomers have even converted measured astronomical signals into audio frequencies, which when played together create an even more literal realization of the harmony of the spheres.

8. Simon Duchesne et al., "Structural and Functional Multiplatform MRI Series of a Single Human Volunteer Over More Than Fifteen Years," *Nature Scientific Data* 6 (2019): 245.

9. Raymond Damadian, "Tumor Detection by Nuclear Magnetic Resonance," *Science* 171, no. 3976 (1971): 1151–53.

10. Not everything Lauterbur predicted in his seminal paper would come to pass. He tried, and fortunately failed, to name his new technique "zeugmatography," from the Greek *zeugma*, meaning "that which is used for joining." Needless to say, his preferred name did not take.

11. Jens Frahm et al., "Rapid NMR Imaging of Dynamic Processes Using the FLASH Technique," *Magnetic Resonance in Medicine* 3, no. 2 (1986): 321–27.

12. Jürgen Hennig et al., "RARE Imaging: A Fast Imaging Method for Clinical MR," *Magnetic Resonance in Medicine* 3, no. 6 (1986): 823–33.

13. In fact, the FLASH and RARE techniques take advantage of two distinct mechanisms of signal decay known to MR scientists as T1 relaxation and T2 relaxation, respectively. T1 relaxation refers to the processes that restore spins to their equilibrium state, whereas T2 relaxation encompasses processes that cause spins to get out of sync with one another so that their individual signals begin to cancel one another out. Since a detailed explication of T1 and T2 processes would be anything but relaxing to some readers, I will simply refer interested readers to extensive online and print tutorials on magnetic resonance phenomena.

14. Denis Le Bihan et al., "MR Imaging of Intravoxel Incoherent Motions: Application to Diffusion and Perfusion in Neurologic Disorders," *Radiology* 161, no. 2 (1986): 401–407.

15. As my colleagues and noted microstructural imaging experts Dmitry Novikov and Els Fieremans are fond of saying, proper interpretation of diffusion contrast can convert an MRI machine into an "in vivo microscope."

16. Chrit T. W. Moonen et al., "Functional Magnetic Resonance Imaging in Medicine and Physiology," *Science* 250, no. 4977 (1990): 53–61.

17. John W. Belliveau et al., "Functional Mapping of the Human Visual Cortex by Magnetic Resonance Imaging," *Science* 254, no. 5032 (1991): 716–19.

18. Seiji Ogawa et al., "Brain Magnetic Resonance Imaging with Contrast Dependent on Blood Oxygenation," *Proceedings of the National Academy of Sciences* 87, no. 24 (1990): 9869–72.

19. Kenneth K. Kwong et al., "Dynamic Magnetic Resonance Imaging of Human Brain Activity During Primary Sensory Stimulation," *Proceedings of the National Academy of Sciences* 89, no. 12 (1992): 5675–79; Seiji Ogawa et al., "Intrinsic Signal Changes Accompanying Sensory Stimulation: Functional Brain Mapping with Magnetic Resonance Imaging," *Proceedings of the National Academy of Sciences* 89, no. 13 (1992): 5951–55; Peter A. Bandettini et al., "Time Course EPI of Human Brain Function During Task Activation," *Magnetic Resonance in Medicine* 25, no. 2 (1992): 390–97.

## 6. PUSHING THE LIMITS

1. Heisenberg's famous uncertainty principle actually represents a kind of diffraction limit for the waves associated with quantum particles.
2. This limit is reflected in Abbe's "numerical aperture," which is defined in terms of opening *angle* rather than diameter, and which reaches a maximum when all available light is collected.
3. De Broglie's Nobel Prize in Physics arrived in 1929, just five years after he completed his doctorate. At the time, he was just thirty-seven years old.
4. The 2017 laureates for cryo-electron microscopy were Jacques Dubochet, Joachim Frank, and Richard Henderson. Their techniques ultimately allowed the visualization of structures at an atomic resolution similar to that of x-ray crystallography, but in frozen samples that did not require the complex and labor-intensive process of crystallization. The 2017 chemistry Nobel citation begins, "A picture is key to understanding. Scientific breakthroughs often build upon the successful visualization of objects invisible to the human eye." This statement certainly holds true for a wide range of imaging advances. Royal Swedish Academy of Sciences, "Press Release: The Nobel Prize in Chemistry 2017," NobelPrize.org, October 4, 2017, https://www.nobelprize.org/prizes/chemistry/2017/press-release/.
5. Another type of close-contact scanning microscope that uses mechanical rather than electronic forces as a means of probing surfaces with a tiny needle tip came to be known as an atomic force microscope.
6. The invention of the laser generated yet another chain of Nobel Prizes, and its provenance was yet again the subject of some dispute.
7. The angular resolution of telescopes is often characterized by the "Rayleigh criterion," named after the nineteenth-century physicist Lord Rayleigh, who published an influential article titled "Resolving, or Separating Power, of Optical Instruments" in 1879. Though Rayleigh came by his criterion through a derivation that differed from what Abbe had used, the two limits amount to much the same thing.
8. At this point, you may be wondering why microscope optics did not grow to similarly impressive sizes. The answer is something I alluded to earlier: microscopes can operate in close proximity to objects of interest, whereas telescopes must be able to operate at a tremendous distance from the objects they are used to scrutinize. Remember that, for nearby small objects, widening a lens opening provides no additional resolution benefits when all the light emerging from those objects is captured. For distant objects, on the other hand, every bit of extra lens or mirror diameter widens the cone of captured light, providing benefits for both resolution and sensitivity.
9. You may have noticed that I have not provided a list of Nobel laureates in telescope technology akin to the list that punctuated the development of modern microscopy. Remarkably few developments in the technology of

astronomical observation have been recognized with Nobel Prizes, though various discoveries using those tools have received recognition. This may in part be a consequence of the bigger-is-better trend that has defined modern astronomy. Today's astronomical instruments require huge teams and significant resources to construct and operate, whereas the Nobel Prize by design singles out only a few seminal contributors. Then again, the Nobel void for astronomical imaging may also reflect human and disciplinary biases that all too often attend high-profile awards.

10. In an exception that proves the rule of Nobel drought in astronomical technology, Martin Ryle did manage to secure a physics Nobel in 1974 for his work in synthesizing large radio telescopes from smaller ones. He received the prize along with Antony Hewish, who used radio telescopes to discover pulsars.

11. The VLA is the source of the radio image of the Crab Nebula in figure 5.4.

12. Or at least the radio emissions emerging from the black hole's outer rim.

13. Astronomers are already making plans to extend the EHT to space via an ambitious mission called the Black Hole Explorer (BHEX). Success of the BHEX will rest on a number of technological breakthroughs, including the development of an ultra-high-speed downlink to bring signals gathered in space down to earth. When it comes to ultra-high-resolution radio astronomy, signal synchronization is no joke. At present, signals from the various earthbound stations in the EHT collective are stored locally on disk, then transported physically to a central location for analysis. Physical transportation is necessary given current limits on the bandwidth and precision of signal transmission from site to site. State-of-the-art GPS systems, atomic clocks, and specialized supercomputers called correlators are called for to get the timings right. Michael D. Johnson et al., "The Black Hole Explorer: Motivation and Vision," arXiv:2406.12917v1 [astro-ph.IM] (2024), https://arxiv.org/abs/2406.12917; "Technology," Event Horizon Telescope, accessed April 14, 2025, https://eventhorizontelescope.org/technology.

14. Ridiculously-long-baseline interferometry (RLBI), anyone? You heard it here . . .

15. In addition to pushing the limits of raw resolution, strong gradients also enable advanced diffusion imaging, which is key for both connectivity mapping and microstructural imaging. The ability to switch between gradient strengths and directions quickly, meanwhile, is a key driver of imaging speed (see chapter 8).

16. Thanks to some clever technological advances, such bulky shielding is no longer needed nowadays, but 7 Tesla scanners remain undeniably hefty.

17. If one were really motivated to push the limits, one could also take a cue from scanning microscopes and put a magnetic resonance probe on the tip of a needle—and, naturally, motivated people have actually done this. The resulting device is no longer really an MRI machine, though. On the other hand, as you heard in chapter 5, imaging approaches like diffusion MRI can be made sensitive to cellular-scale structures without actually resolving them.

18. Some quick ALMA facts, in case you're interested: the ALMA array includes sixty-six antennas, twelve of which are actually mobile. They can be hauled on special transporters to create different antenna configurations. The whole assembly is located in northern Chile's Atacama Desert, whose dry climate and 16,500-foot elevation allow high-frequency radio waves to be captured before they are absorbed by water vapor in the atmosphere. ALMA was built by a consortium of teams from North America, East Asia, and Europe interested in learning about everything from the early universe to newly formed planets.

19. Recall that Peter Mansfield's original formulation of MRI was directly inspired by x-ray crystallography. Meanwhile, subtle properties of magnetic resonance signals, analogous to but different from those exploited in MRI, have been used to map out the structures of free-floating biological molecules too small to see with ordinary light as a direct alternative to using x-ray crystallography to inspect molecules that are difficult to crystallize. The Nobel Prize for this method arrived in 2002, just a year before Lauterbur and Mansfield received their award for macroscopic imaging with magnetic resonance. Connections abound within the extended family of imaging.

20. One notable exception to the rule of isolation among distinct imaging communities may be in the area of image analysis, which not uncommonly brings imagers and images of many different stripes together at meetings centered on cross-cutting computational tools. There is little reason imagers focused on disparate imaging hardware could not follow suit.

## 7. A COMMUNITY OF IMAGERS

1. "ISMRM Strategic Plan," International Society for Magnetic Resonance in Medicine, accessed April 4, 2025, https://www.ismrm.org/strategic_plan.

2. This association, originally known as the American Association of Radiological Technicians, has grown to encompass more than 150,000 members, and it continues to operate today as the American Society of Radiologic Technologists. "History of the American Society of Radiologic Technologists," American Society of Radiologic Technologists, accessed April 5, 2025, https://www.asrt.org/main/about-asrt/museum-and-archives/asrt-history.

3. Various constituents of blood affect the rate of MR signal decay, so the presence of blood can be highlighted by appropriate pulse sequences.

4. For some of Rod's reflections on his experience of racism and the deeper ties that bind us, see Roderic Pettigrew, "Humanity Binds Us," AAMC.org, July 24, 2020, https://www.aamc.org/news/humanity-binds-us.

5. Roderic Pettigrew, "Enter the Physicianeers—How They Will Transform Health Care," *JAMA* 333, no. 8 (2025): 667-68.

6. Jürgen Hennig, "How RARE Came to China: Early Days of MRI," in *eMagRes* (Wiley Online, 2010), https://doi.org/10.1002/9780470034590.emrhp1025.

7. It is a long-standing tradition in the cross-sectional imaging community to produce landmark images of fruits and vegetables. The internal structure of, say, a bell
pepper, can stand in for human anatomy without requiring one to attend to all the
needs and comforts of a human subject. When a 7 Tesla Siemens MRI machine
was installed at NYU in 2004—the second such scanner in the United States and
the first of its kind in the New York area—one of the first images it produced was
a cross-section of an apple. Human heads, knees, and other body parts would follow, but years later, we still show off slices from our "Big Apple" moment.

8. *Medical Imaging Market (2021 Edition)* (Research and Markets, 2021), https://
www.researchandmarkets.com/r/e5u18j.

9. This is how the coils that generate magnetic field gradients look nowadays.

10. C. Douglas Phillips, "RSNA 2022: Disneyland for Radiologists," *Applied Radiology*, November 30, 2022, https://appliedradiology.com/articles/rsna-2022
-disneyland-for-radiologists.

11. Just how much diagnostic imaging drives overall medical costs is a subject of
ongoing debate. A number of recent articles argue that reductions in reim-
bursements for imaging over the last decade or more have actually caused
imaging costs to drop disproportionately while other procedures drive rising
overall costs. See, for example, Monica H. Kassavin et al., "Trends in Medical Part B Payments and Utilization for Imaging Services Between 2009 and
2019," *Current Problems in Diagnostic Radiology* 51, no. 4 (2022): 478–85.

12. "To X-ray or Not to X-ray?," World Health Organization, April 14, 2016, https://
www.who.int/news-room/feature-stories/detail/to-x-ray-or-not-to-x-ray-;
Mahadevappa Mahesh et al., "Patient Exposure from Radiologic and Nuclear
Medicine Procedures in the United States and Worldwide: 2009–2018," *Radiology* 307, no. 1 (2023): e239006; Amy Thompson, "Market Insights: Diagnostic
Imaging Procedure Volumes," LinkedIn, November 30, 2023, https://www
.linkedin.com/pulse/market-insights-diagnostic-imaging-procedure-volumes
-caree.

13. "What Is UK Biobank?," UK Biobank, accessed April 5, 2025, https://www
.ukbiobank.ac.uk.

14. Karla L. Miller et al., "Multimodal Population Brain Imaging in the UK Biobank Prospective Epidemiological Study," *Nature Neuroscience* 19, no. 11 (2016):
1523–36.

15. Gwenaëlle Douaud et al., "SARS-CoV-2 Is Associated with Changes in Brain
Structure in UK Biobank," *Nature* 604, no. 7907 (2022): 697–707.

16. Anahad O'Connor, "Project Baseline Aims to Ward Off Illness Before We Get
Sick," *New York Times*, October 18, 2018.

## 8. IMAGING FOR EVERYONE?

1. Unlike telescopes and microscopes, ordinary cameras have limited zooming
requirements, and their optics can therefore be nicely compact. They do not need
to contend with the diffraction limit to provide highly useful spatial resolution.

2. T. J. Thomson et al., "3.2 Billion Images and 720,000 Hours of Video Are Shared Online Daily. Can You Sort Real from Fake?," *The Conversation*, November 2, 2020, https://theconversation.com/3-2-billion-images-and-720-000-hours-of -video-are-shared-online-daily-can-you-sort-real-from-fake-148630.

3. Matic Broz, "Photo Statistics: How Many Photos Are Taken Every Day?," *Photutorial*, March 2, 2025, https://photutorial.com/photos-statistics.

4. Though the images emerging from the Webb Telescope provide a powerful argument for global connectedness, the telescope's naming has been the subject of controversy. See, for example, Michael Powell, "How Naming the James Webb Telescope Turned into a Fight Over Homophobia," *New York Times*, December 19, 2022.

5. Godwin I. Ogbole et al., "Survey of Magnetic Resonance Imaging Availability in West Africa," *Pan African Medical Journal* 30 (2018): 240–48.

6. Ultrasound has served as a notable exception to the "bigger is better" trend, and I will address this exception later in Chapter 8.

7. Another nontrivial obstacle is that specialized expertise is generally required both to generate and to interpret medical images. Radiographers and radiologists both require specialized training to ply their trade. Once trained, they serve as gatekeepers for their respective imaging devices. Even if medical imaging is nominally for everyone, it cannot at present be done *by* everyone. Unlike for photography, this fact presents a significant practical bottleneck for omnipresence. We will grapple with this challenge in part II.

8. Daniel K. Sodickson and Warren J. Manning, "Simultaneous Acquisition of Spatial Harmonics (SMASH): Fast Imaging with Radiofrequency Coil Arrays," *Magnetic Resonance in Medicine* 38, no. 4 (1997): 591–603.

9. There were of course precedents, as there almost always are in science, but I was blissfully unaware of them when the idea of parallel imaging first occurred to me. As is also quite common in the evolution of scientific understandings or engineering tools, the precedents had gone largely unnoticed until the right seed arrived to crystallize people's attention.

10. Mark went on to become a pioneer not only in parallel imaging, where he introduced a now heavily used technique called GRAPPA (see Mark A. Griswold et al., "Generalized Autocalibrating Partially Parallel Acquisitions (GRAPPA)," *Magnetic Resonance in Medicine* 47, no. 6 (2002): 1202–10), but also in various other creative forms of image acquisition such as magnetic resonance fingerprinting, which will enter our story a little later.

11. Klaas P. Prüssmann et al., "SENSE: Sensitivity Encoding for Fast MRI," *Magnetic Resonance in Medicine* 42, no. 5 (1999): 952–62.

12. Michael Lustig et al., "Sparse MRI: The Application of Compressed Sensing for Rapid MR Imaging," *Magnetic Resonance in Medicine* 58, no. 6 (2007): 1182–95.

13. Liang Gao et al., "Single-Shot Compressed Ultrafast Photography at One Hundred Billion Frames per Second," *Nature* 516 (2014): 74–77.

14. Daniel Kahneman, *Thinking, Fast and Slow* (Farrar, Straus and Giroux, 2011).

15. The use of known physics to guide artificial neural networks has burgeoned in recent years as efforts to offset the "black box" reputation of modern AI have intensified. Two early publications in the area of accelerated MRI are the following: Kerstin Hammernik et al., "Learning a Variational Network for Reconstruction of Accelerated MRI Data," *Magnetic Resonance in Medicine* 79, no. 6 (2018): 3055–71; Hemant K. Aggarwal et al., "MoDL: Model-Based Deep Learning Architecture for Inverse Problems," *IEEE Transactions on Medical Imaging* 38, no. 2 (2019): 394–405. For a comparative clinical evaluation of accelerated AI-reconstructed images, see Michael P. Recht et al., "Using Deep Learning to Accelerate Knee MRI at 3 T: Results of an Interchangeability Study," *American Journal of Roentgenology* 215, no. 6 (2020): 1421–29.

16. In one more nod to Lauterbur and his quixotic attempt to name his new tomographic technique "zeugmatography," the MRI4ALL hackers named their desktop machine the "Zeugmatron Z1." Visit https://github.com/mri4all for more information about the hackathon and its outcomes.

## 9. EMULATING THE SENSES: FROM SNAPSHOTS TO STREAMING

1. Paul Simon, "The Boy in the Bubble," *Graceland*, Warner Brothers Records, 1986.

2. MRI is arguably an extreme example of complex medical imaging workflow, given the sheer quantity of diverse information content we can and often do pack into MR images. But racing against the clock to gather PET projections while a radioactive tracer decays is no picnic either. CT and ultrasound images, whose information content is generally simpler, can be substantially faster to obtain. That said, even for the nominally zippy modality of ultrasound, the operator-dependent manual process of gathering proper cross sections can involve quite a bit of time and effort in practice.

3. Christopher J. Roth et al., "Evaluation of MRI Acquisition Workflow with Lean Six Sigma Method: Case Study of Liver and Knee Examinations," *American Journal of Roentgenology* 195, no. 2 (2010): W150–56. Note that non-value-added time was substantially lower for knee MRI examinations, which have fewer requirements for breath-holding or contrast agent administration.

4. In the case of CT, bad data also constituted a needless addition to a patient's radiation dose.

5. Li Feng et al., "Golden-Angle Radial Sparse Parallel MRI: Combination of Compressed Sensing, Parallel Imaging, and Golden-Angle Radial Sampling for Fast and Flexible Dynamic Volumetric MRI," *Magnetic Resonance in Medicine* 72, no. 3 (2014): 707–17.

6. Anthony G. Christodoulou et al., "Magnetic Resonance Multitasking for Motion-Resolved Quantitative Cardiovascular Imaging," *Nature Biomedical Engineering* 2, no. 4 (2018): 215–26.

7. Dan Ma et al., "Magnetic Resonance Fingerprinting," *Nature* 495, no. 7440 (2013): 187–92.
8. After being captivated by Professor Rucci's lecture to an audience made up largely of neuroscientists, I subsequently invited him to speak before an imaging audience. For those who may be interested, this talk, along with other lectures relevant to the future of imaging, can be found at https://cai2r.net /training/i2i-workshop/i2i-2023/.
9. If you are interested in an accessible introduction to the wonders of how bats use their senses, I refer you once again to Ed Yong's *An Immense World*, which covers echolocation in chapter 9, "A Silent World Shouts Back." It should be noted that in the Introduction to *An Immense World*, Yong cautions against using animals only as a model for our own lives and technologies, and I take his point. Though I am certainly interested in what we, as creatures of imaging, can learn from the way other creatures see and sense the world, my concern is not just technological. Ultimately, I am interested in the miracle of seeing. In fact, I would argue that one goal of artificial imaging is to expand our narrow perceptual experience of the world—what Yong calls our *Umwelt*, following the zoologist Jakob von Uexküll.
10. Ed Yong, *An Immense World: How Animal Senses Reveal the Hidden Realms Around Us* (Random House, 2022), 7.

## 10. EMULATING THE BRAIN: ARTIFICIAL INTELLIGENCE AND THE FUTURE OF IMAGING

1. Feel free to take a quick glance back at figure 4.1 to see just how streaky and unrealistic even a simple image looks with only two projections.
2. In the particular case of actual cars and stoplights, we are justified in driving blithely onward because we know how big a stoplight should look when it is right in front of us.
3. To make the illusion of depth more convincing, we can employ other tricks, like deliberately delivering different views of the same flat surface to our two eyes. This is how 3-D glasses and headsets work. Stereo sound represents a similar sensory illusion: we use multiple recording devices and multiple sound sources to convince our brains that we are hearing voices that aren't really coming from where we think they are.
4. Curtis P. Langlotz, "Will Artificial Intelligence Replace Radiologists?," *Radiology Artificial Intelligence* 1, no. 3 (2019): e190058. The full quote reads as follows: "As we are lifted by the latest AI bubble, 'Will AI replace radiologists?' is the wrong question. The right answer is: Radiologists who use AI will replace radiologists who don't."
5. Eric Topol, *Deep Medicine: How Artificial Intelligence Can Make Healthcare Human Again* (Basic Books, 2019), chapter 6. While my mandate here is to consider the future of seeing, those interested in delving into the many

potential effects of AI on the practice of medicine may want to visit the other chapters of Topol's thought-provoking treatise.

6. Topol notes that, back in 2016, he even made the modest proposal to merge the specialties of radiology and pathology into a unified discipline of imaging experts and information specialists.

7. Perhaps aware of the perils of prediction, Hinton hedged a little on his five-year timeline, adding that "it might be ten years" before deep learning algorithms outperformed radiologists. Then he doubled down, and allowed that "we've got plenty of radiologists anyway.... I said this at a hospital and it didn't go down too well." As of now, it is pretty clear that not many radiologists are likely to be replaced by AI even at the ten-year mark in 2026.

8. Admittedly, this limitation may not last too much longer. AI algorithms are rapidly getting better at understanding context, as I will soon discuss in reference to neural networks like transformers and large language models like ChatGPT.

9. Neil Savage, "Neural Net Worth," *Communications of the ACM* 62, no. 6 (2019): 10–12. Meanwhile, speaking of house cats, some pundits responding to Yann's comments on common sense have noted the irony that the AI algorithms he helped to pioneer have proven to be quite good at identifying online images of cats. Coincidence? You be the judge.

10. Yann LeCun et al., "Deep Learning," *Nature* 521, no. 7553 (2015): 436–44.

11. Purists well schooled in the particulars of backpropagation and back projection might object to this analogy, but I mean it here as a conceptual guide rather than a mathematical identity. We've already invested some time together puzzling over back projection, and I figured it made sense to capitalize on that investment in thinking through the rather abstract backpropagation procedure that is part of the backbone of deep learning.

12. Any attempt to summarize the rich taxonomy of artificial neural networks is guaranteed to be out of date almost immediately, so I will not even try to be complete here.

13. In actual practice, self-driving cars are often equipped with depth-sensitive technology like LIDAR, and/or with multiple sensors that can in principle triangulate depth better than two ordinary eyes. Nevertheless, the problem of automatically sorting out busy 3-D scenes remains a work in progress for the automotive field.

14. Sumit Chopra, "Imaging Without Images: Using AI to Detect Signatures," CAI2R i2i workshop, October 19, 2023, https://cai2r.net/training/i2i-workshop/i2i-2023/.

15. This work put me in an unusual position as a teacher. Instead of training my student to create crisp and clear images, I found myself instructing her in realistic ways to degrade and distort images, in order to see how low we could go in image quality before our neural networks started making mistakes. When my son, Noah, heard this, he grinned widely and said, "Dad, you're evil. You're training imaging weapons!"

16. As was alluded to in chapter 9, moving edges—that is, temporal transitions— are also quick to catch our attention.
17. More common and less whimsical modifications typically include the random removal of selected blocks of pixels, image rotations, changes in size, alterations in lighting, or degradations in signal-to-noise ratio.
18. Self-supervised networks are not necessarily able to provide a familiar name for each class—for that, they would need at least some supervision—but they are able to recognize objects in distinct classes as different from one another.
19. See, for example, Yann LeCun, "Self-Supervised Learning: The Dark Matter of Intelligence," *Meta* (blog), March 4, 2021, https://ai.facebook.com/blog/self -supervised-learning-the-dark-matter-of-intelligence.
20. Bernard Marr, "Generative AI Sucks: Meta's Chief AI Scientist Calls for a Shift to Objective-Driven AI," *Forbes*, April 12, 2024, https://www.forbes.com/sites /bernardmarr/2024/04/12/generative-ai-sucks-metas-chief-ai-scientist-calls -for-a-shift-to-objective-driven-ai/?sh=6ebf312fb82b.
21. Yann LeCun, *A Path Towards Autonomous Machine Intelligence: Version 0.9.2, 2022-06-27* (New York University, June 27, 2022), https://openreview.net /pdf?id=BZ5a1r-kVsf.
22. The same is true, by the way, for many different kinds of computationally intensive simulations. AI appears to have a knack for identifying key features in the behavior of complex systems over time, allowing future behavior to be predicted without resorting to traditional time-consuming and expensive number crunching. In other words, AI can develop a kind of shorthand intuition for physics, getting to the right answer without getting lost in complicated equations. Dramatic recent progress has been shown, for example, in solving long-standing problems in protein folding. In fact, advances in this area have been so dramatic that Demis Hassabis and John Jumper from Deep-Mind shared the 2024 Nobel Prize in Chemistry for their work in developing the AI tool AlphaFold. The third Chemistry laureate that year, David Baker, has noted that his work designing new proteins has also been aided substantially by recent developments in AI.

## 11. NO MORE TUNNEL VISION: IMAGING FOR EVERYONE, EVERYWHERE

1. A recent project by Harvard researchers explored whether smartphone apps and biometric sensors can help to prevent suicide—when social media apps aren't promoting depression, anyway. See Ellen Barry, "Can Smartphones Help Predict Suicide?," *New York Times*, September 30, 2022.
2. Clayton M. Christensen, *The Innovator's Dilemma: When New Technologies Cause Great Firms to Fail* (Harvard Business Review Press, 2015).
3. Frank Preiswerk et al., "Hybrid MRI-Ultrasound acquisitions, and Scannerless Real-Time Imaging," *Magnetic Resonance in Medicine* 78, no. 3 (2017): 897–908.

4. Shiqi Xu et al., "Imaging Dynamics Beneath Turbid Media Via Parallelized Single-Photon Detection," *Advanced Science* 9, no. 24 (2022): e2201885; Leeor Alon, "Electromagnetic Imaging Without the Magnet: Microwave Dielectrography," CAI²R i2i workshop, October 19, 2023, https://cai2r.net/training/i2i-workshop/i2i-2023/.

5. Such intercompatible multisensory representations might be cataloged and compared by something like Yann LeCun's hypothetical world-model engine.

6. "Cancer Facts & Figures 2022," American Cancer Society, accessed April 15, 2025, https://www.cancer.org/research/cancer-facts-statistics/all-cancer-facts-figures/cancer-facts-figures-2022. Those who worry about the world's burgeoning population might take issue with seeking to reduce mortality rates without a comparable reduction in birth rates, but that is a subject for other books to tackle. As a developer of medical technology (rather than, say, cryo-preservation or space colonization technology), I am concerned first and foremost with the quality rather than the quantity of life. Limiting the human and economic costs of prolonged battles with cancer seems to me to be a worthy goal any way you slice it.

7. I serve as a scientific advisor for Ezra, so I cannot comment on its economic standing without conflict of interest. What I can do is explain the basis of my interest.

8. Eric Topol, *Deep Medicine: How Artificial Intelligence Can Make Healthcare Human Again* (Basic Books, 2019), chapter 2.

9. Topol uses the term *digital twin* to refer to another provocative concept: a person whose demographic, biological, and anatomical characteristics closely resemble yours and whose health history may therefore serve as a model for what might happen to you. The availability of dynamic representations of health may in fact allow the robust identification of such resemblances among large collections of people.

10. Dr. Nguyen performed some of this work in collaboration with the late Roger Tsien, who received the 2008 Nobel Prize in Chemistry for developing green fluorescent protein as a powerfully informative dye.

11. So did the crowdsourcing SETI@home initiative, which from 1999 to 2020 pressed home computers into service to analyze signals for signs of extraterrestrial intelligence.

12. Alastair Reynolds, *Poseidon's Wake* (Penguin Random House, 2015), 366.

## 12. THE FUTURE OF SEEING

1. On the legacy of anti-Black discrimination in the field of radiology, see Julia E. Goldberg et al., "How We Got Here: The Legacy of Anti-Black Discrimination in Radiology," *Radiographics* 43, no. 2 (2023): e220112.

2. As was noted at the time, the late Ruth Bader Ginsburg once famously opined that the U.S. Supreme Court would have enough women "when there are

nine." Dave Pearson, "History Pivots: 9 Key Radiology Journals Now Led by Women," *Radiology Business*, January 16, 2023, https://radiologybusiness.com/topics/healthcare-management/leadership/history-pivots-9-key-radiology-journals-now-led-women.

3. Following the arguments from chapter 11, I myself am less interested in the question of whether light can do all this work than in the question of why it needs to. Indirect tomography, multisensor correlation, and learned representations all seem to me like potential enablers of Openwater's ambitious work.

4. I think it is fair to say that finding the distinctive neural signature of an interpretable thought is not as simple as finding a brain region or even an individual cell that might light up on a functional imaging study. As my NYU colleague and noted neuroscientist György Buzsáki explains in his book *The Rhythms of the Brain* (Oxford University Press, 2011), many brain functions involve large populations of brain cells operating in tandem and forming distributed collective modes of activity.

5. Issie Lapowsky, "Why Meta's Yann LeCun Isn't Buying the AI Doomer Narrative," *Fast Company*, September 5, 2023. See also "Yann LeCun, Imagination in Action, Davos 2024," posted February 26, 2024, by Imagination in Action, YouTube, https://www.youtube.com/watch?v=YdaRd_vitLw.

6. As one crude but provocative step on the path to mind reading, researchers have managed to generate approximate replicas of images seen by test subjects by training neural networks to interpret their brain activity, measured with fMRI, while the images were being presented. See Kamal Nahas, "AI Re-creates What People See by Reading Their Brain Scans," *Science News*, March 7, 2023.

7. Tiffany Hsu and Steven Lee Myers, "Can We No Longer Believe Anything We See?," *New York Times*, April 8, 2023.

8. The same was true for a subsequent version, DALL-E 3, which I used to generate selected components of figures 1.2, 2.1 through 2.4, 7.1, and 11.1. The bot produced far prettier pictures than I am capable of drawing, and all that was left for me to do was to assemble the pieces into coherent scenes. Nevertheless, consistent with some of the concerns I expressed in chapter 10, this powerful bot was also surprisingly brittle. Requests for simple modifications of its first responses quickly revealed its lack of an underlying world model in which to locate its creations. Suffice it to say that the laws of physics were seldom respected. As an experiment, I also asked for precise scientific diagrams of the human eye. The result was a ghastly concoction of mismatched parts and confabulated Latin names—worthy of Hieronymus Bosch, perhaps, but by no means suitable for publication. By the time you read this, tools like DALL-E will no doubt have advanced by leaps and bounds, and my early impressions may seem hopelessly quaint. That said, such advances will only further our capacity to be fooled by images.

9. Tom Stoppard, "Night and Day," in *Tom Stoppard: Plays*, vol. 5 (Faber and Faber, 1999), 355. Stoppard later referred to this quote in "Circumspice," his 2013 PEN/Pinter Prize Lecture on bright futures and dark mirrors. The lecture

is reproduced in part in Tom Stoppard, "Tom Stoppard: Information Is Light," *Guardian*, October 11, 2013, https://www.theguardian.com/stage/2013/oct/11 /tom-stoppard-pen-pinter-lecture.

## EPILOGUE: THE CONTINUING STORY OF IMAGING

1. I can think of no more poignant evocation of the power of perspective, and no more suitable epigraph for my final chapter, than the closing words from Zora Neal Hurston's masterpiece, *Their Eyes Were Watching God* (University of Illinois Press, 1978, 286).

2. In an often-paraphrased passage from *Remembrance of Things Past*, Marcel Proust reflects that "the only true voyage of discovery, the only fountain of Eternal Youth, would be not to visit strange lands but to possess other eyes, to behold the universe through the eyes of another, of a hundred others, to behold the hundred universes that each of them beholds, that each of them is." In *An Immense World*, Ed Yong uses this passage as a call to action for understanding the sensory world of animals. It serves just as well as a manifesto for the future of imaging.

3. Arthur Brehmer, *Die Welt in 100 Jahren* [*The World in 100 Years*] (Buntdruck, 1910; repr., Georg Olms Verlag, 2010).

# Bibliography

## BOOKS

Brehmer, Arthur. *Die Welt in 100 Jahren* [*The World in 100 Years*]. Georg Olms Verlag, 2010. Originally published in 1910 by Buntdruck G. mb H.

Buzsáki, György. *The Rhythms of the Brain*. Oxford University Press, 2011.

Chapman, Alan. *England's Leonardo: Robert Hooke and the Seventeenth-Century Scientific Revolution*. CRC Press, 2004.

Christensen, Clayton M. *The Innovator's Dilemma: When New Technologies Cause Great Firms to Fail*. Harvard Business Review Press, 2015.

Darwin, Charles. *On the Origin of Species*. Cassell, 1909 (Google Books reproduction; alternative free edition available via Amazon Classics). Originally published in 1859 by John Murray.

Hurston, Zora Neale. *Their Eyes Were Watching God*. University of Illinois Press, 1978.

Kahneman, Daniel. *Thinking, Fast and Slow*. Farrar, Straus and Giroux, 2011.

Kevles, Bettyann Holtzmann. *Naked to the Bone: Medical Imaging in the Twentieth Century*. Addison-Wesley, 1997.

Land, Michael F., and Dan-Eric Nilsson. *Animal Eyes*. Oxford University Press, 2002.

Reynolds, Alastair. *Poseidon's Wake*. Penguin Random House, 2015.

Stoppard, Tom. "Night and Day." In *Tom Stoppard: Plays*. Vol. 5. Faber and Faber, 1999.

Topol, Eric. *Deep Medicine: How Artificial Intelligence Can Make Healthcare Human Again*. Basic Books, 2019.

Watson, Roger, and Helen Rappaport. *Capturing the Light: The Birth of Photography, a True Story of Genius and Rivalry*. St. Martin's Press, 2013.

Yong, Ed. *An Immense World: How Animal Senses Reveal the Hidden Realms Around Us*. Random House, 2022.

## NEWS ARTICLES AND ESSAYS

Barry, Ellen. "Can Smartphones Help Predict Suicide?" *New York Times*, September 30, 2022.

Dyson, Freeman. "The Scientist as Rebel." *New York Review of Books*, May 25, 1995.

Hsu, Tiffany, and Steven Lee Myers. "Can We No Longer Believe Anything We See?" *New York Times*, April 8, 2023.

Lapowsky, Issie. "Why Meta's Yann LeCun Isn't Buying the AI Doomer Narrative." *Fast Company*, September 5, 2023.

Marr, Bernard. "Generative AI Sucks: Meta's Chief AI Scientist Calls for a Shift to Objective-Driven AI." *Forbes*, April 12, 2024.

Nahas, Kamal. "AI Re-creates What People See by Reading Their Brain Scans." *Science News*, March 7, 2023.

O'Connor, Anahad. "Project Baseline Aims to Ward Off Illness Before We Get Sick." *New York Times*, October 18, 2018.

Powell, Michael. "How Naming the James Webb Telescope Turned into a Fight Over Homophobia." *New York Times*, December 19, 2022.

Savage, Neil. "Neural Net Worth." *Communications of the ACM* 62, no. 6 (2019): 10–12.

Stoppard, Tom. "Tom Stoppard: Information Is Light." *Guardian*, October 11, 2013.

Yong, Ed. "Inside the Eye: Nature's Most Exquisite Creation." *National Geographic*, January 5, 2016.

## SCIENTIFIC AND HISTORICAL ARTICLES

Aggarwal, H. K., M. P. Mani, and M. Jacob. "MoDL: Model-Based Deep Learning Architecture for Inverse Problems." *IEEE Transactions on Medical Imaging* 38, no. 2 (2019): 394–405.

Bandettini, P. A., E. C. Wong, R. S. Hinks, R. S. Tikofsky, and J. S. Hyde. "Time Course EPI of Human Brain Function During Task Activation." *Magnetic Resonance in Medicine* 25, no. 2 (1992): 390–97.

Belliveau, J. W., D. N. Kennedy, R. C. McKinstry, et al. "Functional Mapping of the Human Visual Cortex by Magnetic Resonance Imaging." *Science* 254, no. 5032 (1991): 716–19.

Christodoulou, A. G., J. L. Shaw, C. Nguyen, et al. "Magnetic Resonance Multitasking for Motion-Resolved Quantitative Cardiovascular Imaging." *Nature Biomedical Engineering* 2, no. 4 (2018): 215–26.

Cormack, A. M. "75 Years of Radon Transform." *Journal of Computer Assisted Tomography* 16, no. 5 (1992): 673.

——. "Early Two-Dimensional Reconstruction and Recent Topics Stemming from It." Nobel lecture, December 8, 1979.

Damadian, R. "Tumor Detection by Nuclear Magnetic Resonance." *Science* 171, no. 3976 (1971): 1151–53.

Douaud, G., S. Lee, F. Alfaro-Almagro, et al. "SARS-CoV-2 Is Associated with Changes in Brain Structure in UK Biobank." *Nature* 604, no. 7907 (2022): 697–707.

Dubois, J., M. Alison, S. Counsell, et al. "MRI of the Neonatal Brain: A Review of Methodological Challenges and Neuroscientific Advances." *Journal of Magnetic Resonance Imaging* 53, no. 5 (2021): 1318–43.

Duchesne, S., L. Dieumegarde, I. Chouinard, et al. "Structural and Functional Multiplatform MRI Series of a Single Human Volunteer Over More Than Fifteen Years." *Nature Scientific Data* 6 (2019): 245.

Edelstein, W. A., J. M. Hutchison, G. Johnson, and T. Redpath. "Spin Warp NMR Imaging and Applications to Human Whole-Body Imaging." *Physics in Medicine and Biology* 25, no. 4 (1980): 751–56.

Feng, L., R. Grimm, K. T. Block, et al. "Golden-Angle Radial Sparse Parallel MRI: Combination of Compressed Sensing, Parallel Imaging, and Golden-Angle Radial Sampling for Fast and Flexible Dynamic Volumetric MRI." *Magnetic Resonance in Medicine* 72, no. 3 (2014): 707–17.

Frahm, J., A. Haase, and D. Matthei. "Rapid NMR Imaging of Dynamic Processes Using the FLASH Technique." *Magnetic Resonance in Medicine* 3, no. 2 (1986): 321–27.

Fuchs, V. R., and H. C. Sox Jr. "Physicians' Views of the Relative Importance of Thirty Medical Innovations." *Health Affairs* 20, no. 5 (2001): 30–42.

Gao, L., J. Liang, C. Li, and L. V. Wang. "Single-Shot Compressed Ultrafast Photography at One Hundred Billion Frames per Second." *Nature* 516 (2014): 74–77.

Garcia, D. "SIMUS: an open-source simulator for medical ultrasound imaging. Part I: theory & examples." *Computer Methods and Programs in Biomedicine* 218 (2022): 106726.

Garroway, A. N., P. K. Grannell, and P. Mansfield. "Image Formation in NMR by a Selective Irradiation Process." *Journal of Physics C: Solid State Physics* 7 (1974): L457.

Goldberg, J. E., V. Prabhu, P. N. Smereka, and N. M. Hindman. "How We Got Here: The Legacy of Anti-Black Discrimination in Radiology." *Radiographics* 43, no. 2 (2023): e220112.

Griswold, M. A., P. M. Jakob, R. M. Heidemann, et al. "Generalized Autocalibrating Partially Parallel Acquisitions (GRAPPA)." *Magnetic Resonance in Medicine* 47, no. 6 (2002): 1202–10.

Hammernik, K., T. Klatzer, E. Kobler, et al. "Learning a Variational Network for Reconstruction of Accelerated MRI Data." *Magnetic Resonance in Medicine* 79, no. 6 (2018): 3055–71.

Hennig, J. "How RARE Came to China: Early Days of MRI." In *eMagRes*. Wiley Online, 2010. https://doi.org/10.1002/9780470034590.emrhp1025.

Hennig, J., A. Nauerth, and H. Friedburg. "RARE Imaging: A Fast Imaging Method for Clinical MR." *Magnetic Resonance in Medicine* 3, no. 6 (1986): 823–33.

Hounsfield, G. N. "Computed Medical Imaging." Nobel lecture, December 8, 1979.

——. "Computerised Transverse Axial Scanning (Tomography) I. Description of System." *British Journal of Radiology* 46 (1973): 1016–22.

Johnson, M. D., K. Akiyama, R. Baturin, et al. "The Black Hole Explorer: Motivation and Vision." arXiv:2406.12917v1 [astro-ph.IM] (2024). https://arxiv.org/abs/2406.12917.

Kassavin, M. H., K. D. Parikh, S. H. Tirumani, and N. H. Ramaiya. "Trends in Medical Part B Payments and Utilization for Imaging Services Between 2009 and 2019." *Current Problems in Diagnostic Radiology* 51, no. 4 (2022): 478–85.

Kumar, A., D. Welti, and R. R. Ernst. "NMR Fourier Zeugmatography." *Journal of Magnetic Resonance* 18 (1975): 69–83.

Kwong, K. K., J. W. Belliveau, D. A. Chesler, et al. "Dynamic Magnetic Resonance Imaging of Human Brain Activity During Primary Sensory Stimulation." *Proceedings of the National Academy of Sciences* 89, no. 12 (1992): 5675–79.

Langlotz, C. P. "Will Artificial Intelligence Replace Radiologists?" *Radiology Artificial Intelligence* 1, no. 3 (2019): e190058.

Lauterbur, P. C. "All Science Is Interdisciplinary—From Magnetic Moments to Molecules to Men." Nobel lecture, December 8, 2003.

——. "Image Formation by Induced Local Interactions: Examples Employing Magnetic Resonance." *Nature* 242 (1973): 190–91.

Le Bihan, D., E. Breton, D. Lallemand, P. Grenier, E. Cabanis, and M. Laval-Jeantet. "MR Imaging of Intravoxel Incoherent Motions: Application to Diffusion and Perfusion in Neurologic Disorders." *Radiology* 161, no. 2 (1986): 401–407.

LeCun, Y. *A Path Towards Autonomous Machine Intelligence: Version 0.9.2, 2022-06-27.* New York University, June 27, 2022. https://openreview.net /pdf?id=BZ5a1r-kVsf.

LeCun, Y., Y. Bengio, and G. Hinton. "Deep Learning." *Nature* 521, no. 7553 (2015): 436–44.

Lusebrink, F., A. Sciarra, H. Mattern, R. Yakupov, and O. Speck. "T1-Weighted in Vivo Human Whole Brain MRI Dataset with an Ultrahigh Isotropic Resolution of 250 µm." *Scientific Data* 4 (2017): 170032.

Lustig, M., D. Donaho, and J. M. Pauly. "Sparse MRI: The Application of Compressed Sensing for Rapid MR Imaging." *Magnetic Resonance in Medicine* 58, no. 6 (2007): 1182–95.

Ma, D., V. Gulani, N. Seiberlich, et al. "Magnetic Resonance Fingerprinting." *Nature* 495, no. 7440 (2013): 187–92.

Mahesh, M., A. J. Ansari, and F. A. Mettler Jr. "Patient Exposure from Radiologic and Nuclear Medicine Procedures in the United States and Worldwide: 2009–2018." *Radiology* 307, no. 1 (2023): e239006.

Mallard, J., J. M. S. Hutchison, W. A. Edelstein, et al. "In Vivo N.M.R. Imaging in Medicine: The Aberdeen Approach, Both Physical and Biological." *Philosophical Transactions of the Royal Society of London B* 289, no. 1037 (1980): 519–30.

Mansfield, P., and P. K. Grannell. "NMR 'Diffraction' in Solids?" *Journal of Physics C: Solid State Physics* 6 (1973): L422–26.

Miller, K. L., F. Alfaro-Almagro, N. K. Bangerter, et al. "Multimodal Population Brain Imaging in the UK Biobank Prospective Epidemiological Study." *Nature Neuroscience* 19, no. 11 (2016): 1523–36.

Moonen, C. T. W., P. C. M. van Zijl, J. A. Frank, D. Le Bihan, and E. D. Becker. "Functional Magnetic Resonance Imaging in Medicine and Physiology." *Science* 250, no. 4977 (1990): 53–61.

Nilsson, D.-E., and S. Pelger. "A Pessimistic Estimate of the Time Required for an Eye to Evolve." *Proceedings of the Royal Society B* 256, no. 1345 (1994): 53–58.

Ogawa, S., T. M. Lee, A. R. Kay, and D. W. Tank. "Brain Magnetic Resonance Imaging with Contrast Dependent on Blood Oxygenation." *Proceedings of the National Academy of Sciences* 87, no. 24 (1990): 9869–72.

Ogawa, S., D. W. Tank, R. Menon, et al. "Intrinsic signal Changes Accompanying Sensory Stimulation: Functional Brain Mapping with Magnetic Resonance Imaging." *Proceedings of the National Academy of Sciences* 89, no. 13 (1992): 5951–55.

Ogbole, G. I., A. O. Adeyomoye, A. Badu-Peprah, Y. Mensah, and D. A. Nzeh. "Survey of Magnetic Resonance Imaging Availability in West Africa." *Pan African Medical Journal* 30 (2018): 240–48.

Pettigrew, R. "Enter the Physicianeers—How They Will Transform Health Care." *JAMA* 333, no. 8 (2025): 667-68.

Preiswerk, F., M. Towes, C.-C. Cheng, et al. "Hybrid MRI-Ultrasound Acquisitions, and Scannerless Real-Time Imaging." *Magnetic Resonance in Medicine* 78, no. 3 (2017): 897–908.

Prüssmann, K. P., M. Weiger, M. B. Scheidegger, and P. Bösiger. "SENSE: Sensitivity Encoding for Fast MRI." *Magnetic Resonance in Medicine* 42, no. 5 (1999): 952–62.

Recht, M. P., J. Zbontar, D. K. Sodickson, et al. "Using Deep Learning to Accelerate Knee MRI at 3 T: Results of an Interchangeability Study." *American Journal of Roentgenology* 215, no. 6 (2020): 1421–29.

Roth, C. J., D. T. Boll, L. K. Wall, and E. M. Merkle. "Evaluation of MRI Acquisition Workflow with Lean Six Sigma Method: Case Study of Liver and Knee Examinations." *American Journal of Roentenology* 195, no. 2 (2010): W150–56.

Sodickson, D. K., and W. J. Manning. "Simultaneous Acquisition of Spatial Harmonics (SMASH): Fast Imaging with Radiofrequency Coil Arrays." *Magnetic Resonance in Medicine* 38, no. 4 (1997): 591–603.

Spiegel, Peter K. "The First Clinical X-ray Made in America—100 Years." *American Journal of Roentgenology* 164 (1995): 241–43.

Van Helden, A. "The Invention of the Telescope." *Transactions of the American Philosophical Society* 67, no. 4 (1977): 1–67.

White, D. N. "Neurosonology Pioneers." *Ultrasound in Medicine and Biology* 14, no. 7 (1988): 541–61.

Xu, S., X. Yang, W. Liu, et al. "Imaging Dynamics Beneath Turbid Media Via Parallelized Single-Photon Detection." *Advanced Science* 9, no. 24 (2022): e2201885.

## ONLINE SOURCES

Broz, Matic. "Photo Statistics: How Many Photos Are Taken Every Day?" *Phototorial*, March 2, 2025. https://phototorial.com/photos-statistics.

"Cancer Facts & Figures 2022." American Cancer Society, accessed April 15, 2025. https://www.cancer.org/research/cancer-facts-statistics/all-cancer-facts-figures/cancer-facts-figures-2022.

"Crab Nebula." NASA, accessed April 22, 2023. https://www.nasa.gov/multimedia/imagegallery/image_feature_567.html.

"History of the American Society of Radiologic Technologists." American Society of Radiologic Technologists, accessed April 5, 2025. https://www.asrt.org/main/about-asrt/museum-and-archives/asrt-history.

"i2i Workshop 2023: Meeting Program, Talk Recordings, and Poster PDFs." CAI$^2$R, October 2023. https://cai2r.net/training/i2i-workshop/i2i-2023/.

"ISMRM Strategic Plan." International Society for Magnetic Resonance in Medicine, accessed April 4, 2025. https://www.ismrm.org/strategic_plan.

LeCun, Yann. "Self-Supervised Learning: The Dark Matter of Intelligence." Meta (blog), March 4, 2021. https://ai.facebook.com/blog/self-supervised-learning-the-dark-matter-of-intelligence.

*Medical Imaging Market (2021 Edition)*. Research and Markets, 2021. https://www.researchandmarkets.com/r/e5u18j.

"MRI4ALL Hackathon 2023." MRI4ALL, October 2023. https://github.com/mri4all.

"The Nobel Prize in Physics 1901." NobelPrize.org, accessed March 26, 2025. https://www.nobelprize.org/prizes/physics/1901/summary.

Pearson, Dave. "History Pivots: 9 Key Radiology Journals Now Led by Women." *Radiology Business*, January 16, 2023. https://radiologybusiness.com/topics/healthcare-management/leadership/history-pivots-9-key-radiology-journals-now-led-women.

Pettigrew, Roderic. "Humanity Binds Us." AAMC.org, July 24, 2020. https://www.aamc.org/news-insights/humanity-binds-us.

Phillips, C. Douglas. "RSNA 2022: Disneyland for Radiologists." *AppliedRadiology*, November 30, 2022. https://appliedradiology.com/articles/rsna-2022-disneyland-for-radiologists.

Royal Swedish Academy of Sciences. "Press Release: The Nobel Prize in Chemistry 2017." NobelPrize.org, October 4, 2017. https://www.nobelprize.org/prizes/chemistry/2017/press-release.

Stoppard, Tom. "Tom Stoppard: Information Is Light." *The Guardian*, October 11, 2013. https://www.theguardian.com/stage/2013/oct/11/tom-stoppard-pen-pinter-lecture#:~:text=People%20do%20awful%20things%20to,all%20the%20other%20freedoms%20possible.

"Technology." Event Horizon Telescope, accessed April 14, 2025. https://eventhorizontelescope.org/technology.

Thompson, A. "Market Insights: Diagnostic Imaging Procedure Volumes." *LinkedIn*, November 30, 2023. https://www.linkedin.com/pulse/market-insights-diagnostic-imaging-procedure-volumes-caree.

Thomson, T. J., D. Angus, and P. Dootson. "3.2 Billion Images and 720,000 Hours of Video Are Shared Online Daily. Can You Sort Real from Fake?" *The Conversation*, November 2, 2020. https://theconversation.com/3-2-billion-images-and-720-000-hours-of-video-are-shared-online-daily-can-you-sort-real-from-fake-148630.

"To X-ray or Not to X-ray?" World Health Organization, April 14, 2016. https://www.who.int/news-room/feature-stories/detail/to-x-ray-or-not-to-x-ray-.

"What Is UK Biobank?" UK Biobank, accessed April 5, 2025. https://www.ukbiobank.ac.uk.

"Yann LeCun, Imagination in Action, Davos 2024." Posted February 26, 2024, by Imagination in Action. YouTube. https://www.youtube.com/watch?v=YdaRd_vitLw.

Yong, Ed. "Inside the Eye: Nature's Most Exquisite Creation." *National Geographic*, January 5, 2016. https://www.nationalgeographic.com/magazine/article/evolution-of-eyes.

Zagzebski, James. "History of Ultrasound Imaging." 54th Annual Meeting of the American Association for Physics in Medicine, accessed March 30, 2025. https://www.aapm.org/education/vl/vl.asp?id=397.

# Index

Page numbers in *italics* refer to images.